The Rightful Place of Science:

Disasters & Climate Change

The Rightful Place of Science:

Disasters & Climate Change

Roger Pielke, Jr.

Foreword by
Daniel Sarewitz

Consortium for Science, Policy & Outcomes
Tempe, AZ and Washington, DC

For information on the Rightful Place of Science series,
write to: Consortium for Science, Policy & Outcomes
PO Box 875603, Tempe, AZ 85287-5603

Model citation for this volume:

Pielke, Jr., R. 2014. *The Rightful Place of Science: Disasters & Climate Change*. Tempe, AZ: Consortium for Science, Policy & Outcomes.

The Rightful Place of Science series explores the complex interactions among science, technology, politics, and the human condition.

Other volumes in this series:

Sarewitz, D., ed. 2014. *The Rightful Place of Science: Government & Energy Innovation*. Tempe, AZ: Consortium for Science, Policy & Outcomes.

Alic, John A. 2013. *The Rightful Place of Science: Biofuels*. Tempe, AZ: Consortium for Science, Policy & Outcomes.

Zachary, G.P., ed. 2013. *The Rightful Place of Science: Politics*. Tempe, AZ: Consortium for Science, Policy & Outcomes.

ISBN: 0692297510

ISBN-13: 978-0692297513

LCCN: 2014917724

FIRST EDITION, NOVEMBER 2014

CONTENTS

FOREWORD

Daniel Sarewitz[*]

Effective action in the world — getting done what we want to get done — depends on three fundamental things: a coherent, shared vision of what we want to accomplish; an accurate understanding of the current conditions for taking action; and, consistent with that understanding, a practical approach to pursuing our goals. In the real world this is often much easier said than done, especially when a problem is complicated.

Consider for a moment the U.S. invasion of Iraq in 2003. The goals of the invasion were clear enough — to remove Saddam Hussein from power, eliminate Iraq's supposed capacity to produce weapons of mass destruction, and create the conditions for a democratic regime to emerge. But the situation in Iraq was greatly misunderstood by the Bush Administration and the U.S. intelligence agencies, and the approach to action was thus entirely inappropriate to the real complexity of the situa-

---

[*] Daniel Sarewitz is co-director of the Consortium for Science, Policy & Outcomes, and Professor of Science and Society at Arizona State University.

tion. We continue to pay the consequences of this bad decision.

The situation with disasters and climate change would appear to be much the opposite. The costs of disasters are rising. This is a moral and economic challenge for all societies. Almost everyone would agree that the goal of reducing societal vulnerability to disasters caused by hurricanes, floods, droughts, and other climate-related hazards is worth pursuing. As well, the relevant facts about disasters and climate change are actually quite clear and scientifically uncontroversial.

In this book, Roger Pielke, Jr. summarizes those facts to answer the question, "Have disasters become more costly because of human-caused climate change?" Many people do worry that climate change is causing disasters to get worse, but Pielke presents a wealth of data, including the conclusions of the Intergovernmental Panel on Climate Change, to show why such concerns are not supported by the available science. Unlike the conditions in Iraq prior to the U.S. invasion, the reasons for rising disaster losses are well understood and unlikely to change significantly with new revelations or data.

Why, then, are disaster costs rising? The reasons are apparent: populations continue to grow, the economy and the built environment continue to expand, people migrate to and concentrate on coastal and flood plains. There are simply more people, and more of the things that people depend on in their lives, in harm's way. Moreover, these demographic trends feed continued environmental degradation of highly populated coastal, riverine, and mountainous regions, which in turn exacerbate the consequences of disasters. Most of these trends are further amplified in developing countries. Climate isn't the only thing that's changing in our world, and it's these other changes that are causing disaster losses to increase.

The only politically and practically feasible way to slow this increase, let alone stabilize or even reverse it, is to improve societal preparedness. When floods devastated the Netherlands in 1953, the nation came together to devise institutions, policies, and projects that would prevent such a catastrophe from happening again. Back then no one worried about whether climate change contributed to the disaster; much of the nation was already built below sea level. Proven policy tools for reducing disaster vulnerability include public education, better (and better-enforced) building codes and land-use practices, improved infrastructure, sensible insurance programs, enhanced warning, emergency planning and response capabilities, and so on. By deploying such tools, many places in the world, including poor countries, have made great progress on reducing their vulnerability to disasters.

Disasters are a serious problem, as are human-caused changes to our climate. Taking them both seriously, and addressing them effectively, requires the recognition that they are not serious for the same reasons, and that the pathways for addressing them are different, and must respond to different information, arguments, motives, and policies. Reducing greenhouse gas emissions remains an urgent policy imperative, but one that will have no capacity to reduce disaster losses in the foreseeable future, and will never be rigorously justifiable in terms of measurable reductions in disaster vulnerability. Another way to think about this problem is that even if climate change wasn't happening, or suddenly ceased, all of the factors that are causing disaster loss increases would still be as powerful as ever.

Decades hence, climate change may well play a discernible role in making disasters worse, but even then, the moral and practical imperative is to reduce losses, regardless of their cause, by directly acting to improve disaster preparedness in all societies. This book shows that the moral and practical perspectives are also backed by another powerful motivator of effective action: the science. Rarely is the case for effective action so clear.

INTRODUCTION

This short volume is for those interested in understanding the science of disasters and climate change.

I have three motivations for writing it.

First, while the Intergovernmental Panel on Climate Change has done excellent work in recent years to summarize a vast literature on this subject, the resulting assessments of the current state of science span thousands of pages across dozens of chapters in multiple volumes. Few people will have the time or expertise to sort through the work of the IPCC and arrive at a concise summary of the state of science on disasters and climate change. This short volume provides such a synthesis of the current state of the science.

The science presented in this volume is my own synthesis of the state of the science, and it is fully consistent with the work of the IPCC, but is somewhat more detailed and focused. To underscore the degree of congruence between my synthesis and the views of the IPCC, in several sections you will find extended quotes from the IPCC, rather than reinterpretations. This is to facilitate accurate representation of the IPCC's conclusions. Of course, for those who want more depth, I encourage you to dive into the IPCC reports and especially the primary literature which it draws on, and also those new studies which have been published more recently. One of the wonderful things about science is that it constantly evolves. Any assessment is just a snapshot in time. That goes for the work of the IPCC as well as this short book.

Second, as an academic who has studied and written about disasters and climate change since my days as a post-doc at the National Center for Atmospheric Research in the early 1990s, I am often called on—by governments, industry, the media—to share my views on this subject. But because climate change is a deeply politicized issue, and disasters are at the forefront of the debate, regrettably at times my views have been misrepresented by those seeking to delegitimize them.

Part of this may be my own fault as my work is more sprawling than the IPCC's, appearing in various books, academic journals, blog posts, tweets, and commentaries over a period of almost 20 years. This book provides my views on disasters and climate change between two covers. The final chapter offers some of my views on the policy and politics of the climate debate, but a better source for that information is my book, *The Climate Fix* (Basic Books, 2011).

The third motivation for this book is the degree to which the science of disasters and climate change has become so politicized. I decided that this book was necessary when in the spring of 2014 I saw on the website of the White House a claim that in the United States floods and drought have become more common. Actually, the scientific assessment which the White House produced and then relied on to make these claims says that they have not.[1]

---

[1] The claims of increasing floods and drought are found at "President Obama's Plan to Fight Climate Change," The White House website (25 June 2013), available at:
http://www.whitehouse.gov/share/climate-action-plan; the

While political spin doctors often find ways to parse language and statistics to say things plausibly defensible, but which are ultimately misleading or just wrong, for me this went too far in an area where I have some considerable expertise. The willingness of some in the media and the scientific community to let such claims stand uncorrected for reasons of political expediency does not offer a route to scientific integrity. As you'll see in the pages below, at some point in my career I decided that on topics where I have expertise, I have an obligation to participate in public debates.

So while I may continue to write scientific papers, blog posts, tweets, and commentaries on the subject as I have in the past, I've now also written this short volume. It represents my contribution to upholding scientific integrity in the climate debate. By reading it, you'll have a better sense of the state of the science on disasters and climate change as it stands in 2014, and be in a better position to assess some of the claims being made in the ongoing debate about the societal impacts of human-caused climate change.

Roger Pielke, Jr.
Boulder, Colorado
October, 2014

National Climate Assessment on which the claims are based says: "There has been no universal trend in the overall extent of drought across the continental U.S. since 1900" and "when averaging over the entire contiguous U.S., there is no overall trend in flood magnitudes." The assessment report is available at: http://nca2014.globalchange.gov/

1

CLIMATE'S LEGITIMACY WARS

The earth's climate system is the basis for all life on the planet. It creates the conditions which allow humans and the ecosystems on which we all depend to flourish. Hence it is not surprising that how we are changing that climate system, as a by-product of energy consumption and other factors, attracts considerable interests and passions. People have strong views as to what others should think and do about climate change. Consequently, climate change is hotly contested and vigorously debated in the political arena.

The fact that the issue of climate change is political is not a problem. Politics is how we manage the business of living together. We bargain, negotiate, and compromise, and oftentimes that process is not very pretty. Climate change politics are of course no different. Anyone who wants to participate in the very public and very intense debate over climate change should be ready for some sharp elbows. As Harry S. Truman once said, "If you can't stand the heat, get out of the kitchen."

I've been cooking in the climate kitchen for a long time. One subject, which has mostly been on the back burner but for which the heat has been turned up in recent years, is the relationship of human-caused climate change and disasters.

In recent years, advocates for action on climate change have enlisted disasters as a leading theme of advocacy campaigns, ultimately focused on motivating political action on energy policy. A turn to this strategy has occurred despite a broad consensus in the scientific literature that the evidence for connections between climate change and disasters is incredibly weak, as reflected in the 2012, 2013, and 2014 reports of the Intergovernmental Panel on Climate Change (IPCC), the United Nations (UN) body formed in 1988 to periodically assess the state of climate science.

More specifically, disasters have become both more costly and less deadly over the past century. But there is precious little evidence to suggest that the blame for the increasing tally of disaster costs can be placed on more frequent or extreme weather events attributable to human-caused climate change.

This is an important conclusion because it tells us that the disasters that we experience are largely a consequence of decisions that we make — where we locate our communities, how we build them, how we prepare for the future, and so on. As Gilbert White, the great geographer and disasters expert, wrote in 1945, "Floods are 'acts of God,' but flood losses are largely acts of man."[1] But in an era of climate change, disasters, including floods, may be more than just "acts of God." It is not unreasonable to surmise that we may indeed be influencing the frequency and intensity of extreme events.

Science is useful because it allows us to do more than just surmise. We can look at evidence, compare it with theory, and make a judgment as to whether we can detect any influence of changes in climate on the disasters that

---

[1] "Selected Quotations of Gilbert F. White," Natural Hazards Center, University of Colorado at Boulder website, available at: http://www.colorado.edu/hazards/gfw/quotes.html

we experience. So far at least, the data don't support claims that we can identify that influence with respect to those extremes which cause the most damage to property.

In fact, based on the current expectations of the climate science community — specifically that humans are impacting the climate and that these impacts will become more significant in the future as projected by the IPCC — there is presently very little basis for expecting that changes in climate will lead to a demonstrable increase in the costs of disasters any time soon. It will likely be many decades before such a signal can be detected in disaster losses based on current scientific understandings.

In this context, it is surprising that many champions of action on climate change are basing their campaigns on strong claims that are at odds with the current state of scientific knowledge. That's just not smart politics. And it's not just the more aggressive environmental groups that have jumped on the bandwagon linking disasters to climate change.

For instance, in a June, 2013 radio address President Barack Obama explicitly linked disasters and climate change:[2]

> [W]hile we know no single weather event is caused solely by climate change, we also know that in a world that's getting warmer than it used to be, all weather events are affected by it — more extreme droughts, floods, wildfires, and hurricanes.
>
> Those who already feel the effects of a changing climate don't have time to deny it — they're busy dealing with it. The fire-

---

[2] In May, 2014 the White House upped the ante by organizing its climate policy advocacy around extreme events. See "Climate Change and President Obama's Action Plan," The White House website, available at: http://www.whitehouse.gov/climate-change

fighters who brave longer wildfire seasons. The farmers who see crops wilted one year, and washed away the next. Western families worried about water that's drying up.

The cost of these events can be measured in lost lives and livelihoods, lost homes and businesses, and hundreds of billions of dollars in emergency services and disaster relief. And Americans across the country are already paying the price of inaction in higher food costs, insurance premiums, and the tab for rebuilding."[3]

Having studied disasters and climate change for 20 years and having published dozens of papers on the subject, when I heard the president make these remarks I knew that several (but not all) of the claims he made were just plain wrong—they were not supported by the state of the research. In fact, some were contradicted by that research.[4]

When a prominent public official makes claims that are wrong on a subject that an academic has considerable expertise in does one speak out and try to expose the official's mistakes, but risk becoming embroiled in a political debate? Should one just stay silent, maintaining a dignified academic distance, but also maintaining irrelevance? Context matters, of course, and different people will have

---

[3] "Weekly Address: Confronting the Growing Threat of Climate Change," The White House, Office of the Press Secretary press release (29 June 2013), available at: http://www.whitehouse.gov/the-press-office/2013/06/29/weekly-address-confronting-growing-threat-climate-change; as will be documented in detail in the pages that follow, the U.S. has not experienced an increase in hurricanes, floods, or drought, according to the Obama Administration's own National Climate Assessment.

[4] Most obviously, there is no evidence of "more hurricanes" in the U.S. or globally, as documented in detail below. The president's claims about droughts and floods are also questionable, as explained in the previous footnote.

different views on this subject. There are appropriate and legitimate reasons for those different views. Of course, partisans in a political debate whose agendas are affected by the public claims being advanced also have strong views as to whether academics should speak out, usually determined as a matter of political expediency.

Speaking in 2013, former Vice President Al Gore explained the political importance of tying extreme events to climate change in the campaign for action, as reported by *The Hill*:

> *Gore said there's a political interest in determining climate change causes extreme weather. He said lawmakers cannot address the root of disasters without first making a connection between emissions, climate change and extreme weather.*
>
> *Failing to acknowledge that connection will imperil future relief efforts as disasters grow more frequent and expensive, Gore said... Gore advocated putting a price on carbon to limit emissions as a way to subdue those incidents.*[5]

The explicit connection of climate change to extreme weather to putting a price on carbon has the effect of fusing these issues together in the political debate. In such a context it is consequently easy to conflate (a) belief in the connection of climate change and extremes with (b) support for a specific political prescription—in this case, a price on carbon in order to reduce the impacts of extremes. This is a great example of how science can become deeply and irrevocably politicized.

Having written a book calling for a price on carbon *and* for greater scientific integrity with respect to the science of

[5] Z. Colman, "Gore laments scientists 'won't let us' tie climate change to tornadoes," *The Hill* (11 June 2013), available at: http://thehill.com/blogs/e2-wire/e2-wire/304755-gore-says-record-breaking-tornadoes-a-result-of-climate-change

extreme events, I knew that the issue of disasters and climate change did not have to be fused together in political debate.

Climate change and disasters have not always been so central to the politics of the issue.

In 2006 my work on disasters and climate change was viewed to be sufficiently notable that I was invited by the Ocean Sciences Board of the National Academy of Sciences to give its Roger Revelle commemorative lecture at the Natural History Museum of the Smithsonian Institution.[6] Revelle was the scientist who first captured Al Gore's attention on the issue of human-caused climate change.

Following a delightful dinner among the dinosaurs in the museum,[7] I gave a talk in which I explained:

> *To emphasize, humans have an effect on the global climate system and reducing greenhouse gas emissions makes good sense. But reducing emissions will not discernibly affect the trend of escalating disaster losses because the cause of that increase lies in ever-growing societal vulnerability.[8]*

In early 2006, the science of the role of climate change in extreme weather and disasters was not widely viewed as central to the broader political debate over climate policies.

---

[6] "Roger Revelle Lecture Series," National Academy of Sciences website, available at: http://nas-sites.org/revellelecture/past-lecturers/2006-2/

[7] Here I reference the actual fossils, not members of the NAS.

[8] R. Pielke, Jr., "Disasters, Death, and Destruction: Accounting for Recent Calamities," Seventh Annual Roger Revelle Commemorative Lecture (15 Mar. 2006), available at: http://nas-sites.org/revellelecture/files/2011/11/Revelle_program_2006.pdf

In just a few years, however, extreme weather would move to the center of the entire debate over climate change. Figure 1.1 illustrates the increasing presence of the phrase "extreme weather" on the pages of the *New York Times*. From 2006 to 2013 the use of the phrase increased by a factor of 10. As the politics heated up surrounding this issue, any suggestion that human-caused climate change was not driving the rapid increase in disaster costs came to be viewed by some participants in the debate as illegitimate, even heretical. I've had a front row seat to that transformation.

Figure 1.1: Number of Articles Mentioning "Extreme Weather" in the *New York Times* (1965-2014)

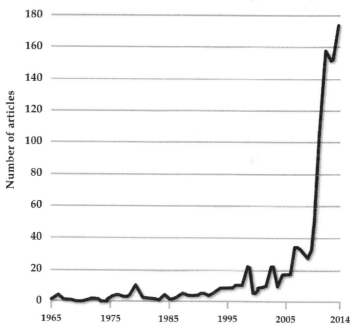

Source: Data from the Chronicle search tool of the *New York Times*, accessed 20 July 2014, available at: http://chronicle.nytlabs.com

My views on what role I should play in the public debate over climate and extreme events have been shaped by my experiences. Back in 2001, soon after George W. Bush was elected president, I was invited by the U.S. National Academy of Sciences to join a small group of experts to brief several senators and the new Secretary of Treasury, Paul O'Neill, on various aspects of climate science. Soon after the event was announced publicly, much to my surprise, I found myself being lobbied by some of my scientific colleagues about what I should say to the policy makers.

My surprise was not that I was being lobbied, as this was a pretty influential forum for any academic. My surprise was that my colleagues were asking me to downplay and to even misrepresent my own research because it was viewed as being inconvenient in the advocacy effort on climate change. My work had found no evidence of a signal of human-caused climate change in the growing toll of losses from floods, hurricanes, and other extremes. While I had concluded that actions to reduce emissions of greenhouse gases made good sense, I also believed that pointing to the latest disasters in advocacy for action went beyond what the science could support, and thus should be avoided.

At the time I likened the pressure from a few of my peers to the following hypothetical:

Imagine that as policy makers are debating intervening militarily in a foreign country, the media report that 1,000 women and children were brutally murdered in that country. This report inflames passions and provides a very compelling justification for the military intervention. A journalist discovers that, contrary to the earlier reports, only 10 soldiers died. What is the journalist's obligation to report the "truth" knowing full well that it might affect political sentiments that were shaped by the earlier erroneous report?

> *When science is used (and misused) in political advocacy,*
> *there are frequent opportunities for such situations to arise.* [9]

Of course, September 11, 2001 occurred just a few months after my meeting with Secretary O'Neill, and a version of this hypothetical became very real with Vice President Dick Cheney, among others, linking that terrorist attack to Saddam Hussein, and using this linkage in public statements as a reason why the public should support an invasion of Iraq. [10]

Cheney later admitted that those claims of attribution were wrong, but waved them away as irrelevant because they supported the important goal of removing Hussein from power. The ends, he seemed to imply, justified the means. From the standpoint of military intelligence, in the United Kingdom, Robin Cook, a member of Tony Blair's cabinet who resigned his position in March 2003 in protest over Iraq, explained: "Instead of using intelligence as evidence on which to base a decision about policy, we used intelligence as the basis on which to justify a policy on which we had already settled." [11]

Irrespective of the merits of the policy, using information in this way undermines scientific integrity. My own experience with being pressured to downplay the conclusions of my research, along with misuse of evidence

---

[9] R. Pielke, Jr., "Editorial: Reflections on Science and Policy," *WeatherZine* 28 (June 2001), available at: http://sciencepolicy.colorado.edu/zine/archives/1-29/txt/zine28.txt

[10] "Cheney: No link between Saddam Hussein, 9/11," CNN Politics website (1 June 2009), available at: http://www.cnn.com/2009/POLITICS/06/01/cheney.speech/

[11] W. Hoge, "2 Former Cabinet Members Say Britain Exaggerated Iraq Claims," *New York Times* (17 June 2003), available at: http://www.nytimes.com/2003/06/17/international/europe/17CND-BRIT.html

in the run-up to the Iraq war, shaped my views and my commitment to speak out on topics where I have research expertise. To avoid science serving as a proxy for debates over politics, it is important when speaking out to be able to place science into a policy context.

To help explain to the scientific community the dangers of ends-justify-means uses of expertise, I included a case study of the pathologies of the role of intelligence in the Iraq War decision in my 2007 book on science and politics, *The Honest Broker* (Cambridge University Press, 2007).

What I experienced in 2001 and what we saw in the build-up to the Iraq War have been described by political scientist Aynsley Kellow as "noble cause" corruptions of science. Kellow explains:

> *The good cause — one that most of us support — can all too readily corrupt the conduct of science, especially science informing public policy, because we prefer answers that support our political preferences, and find science that challenges them less comfortable.[12]*

The issue of disasters and climate change is a canonical example of "noble cause" corruption in science. Such corruption even found its way into the authoritative reports of the IPCC back in 2007.

Noble Cause Corruption in IPCC 2007

In 2007 the IPCC found itself embroiled in a bit of controversy. This controversy had nothing to do with the core findings of climate research related to the physical science, impacts, or economics of climate change. Rather it had to

---

[12] A. Kellow, "All in a good cause," *Online Opinion* (16 May 2008), available at: http://www.onlineopinion.com.au/view.asp?article=7368

do with a few claims in the report which had been exaggerated or were mistaken, as well as the panel's excessive reliance on non-peer reviewed sources, and its public stance of arrogant resistance to owning up to clear errors.

While many people have heard about the erroneous projection in the report that Himalayan glaciers would melt away by 2035, there was a far more egregious error in my area of expertise.

When the 2007 IPCC report came out I was surprised to see that it had included in its assessment a graph (reproduced in Figure 1.2) which showed increasing disaster losses plotted alongside increasing global temperatures, with the vertical scale jiggered to make them appear to increase in lockstep. I was surprised because I had never seen any such graph in the scientific literature. How had I missed something this important?

Figure 1.2: The "Mystery Graph"

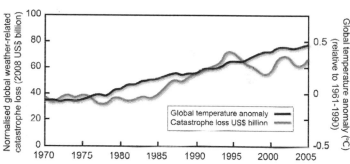

Source: M.L. Parry, O.F. Canziani, J.P. Palutikof, P.J. van der Linden, and C.E. Hanson, eds., *Contribution of Working Group II to the Fourth Assessment Report of the Intergovernmental Panel on Climate Change* (Cambridge, UK: Cambridge University Press, 2007).

My surprise quickly doubled when I saw that the graph was referenced to a non-peer-reviewed white paper that *I had commissioned* for a 2006 workshop on climate change and disasters, which I had organized in collaboration with Peter Höppe of Munich Reinsurance.[13] I knew that paper had no such graph or analysis. So where it actually came from was a complete mystery.

Despite my sleuthing efforts, chronicled in public on my blog, the origins of that "mystery graph" persisted for almost three years. Then in February, 2010 on a cold, damp evening at the historical Royal Institution in London, I participated in a "debate" on disasters and climate change. (I put "debate" in scare quotes because it did not wind up being much of a debate, with all three participants essentially in complete agreement on the state of the science in his area.)

One participant that evening was Robert Muir-Wood, an employee of the catastrophe modeling firm RMS and also a lead author of the 2007 IPCC report.[14]

At the debate Muir-Wood made a remarkable admission. He had been the one who created that graph—"informally" in his words. With the advantage of hindsight he said, "Personally, I think that it should not have been there."[15] A few days later his company, RMS, issued a press release amplifying the point: "RMS believes that

---

[13] P. Höppe and R. Pielke, Jr., eds., *Workshop on Climate Change and Disaster Losses: Understanding and Attributing Trends and Projections*, Final Workshop Report (Hohenkammer, Germany: Munich Re and University of Colorado, 25-26 May 2006), available at:
http://cstpr.colorado.edu/sparc/research/projects/extreme_events/munich_workshop/workshop_report.html

[14] The third participant that evening was Bob Ward, a public relations specialist from the London School of Economics.

[15] R. Pielke, Jr., "IPCC Mystery Graph Solved!" Roger Pielke Jr.'s Blog (8 Feb. 2010), available at:
http://rogerpielkejr.blogspot.com/2010/02/ipcc-mystery-graph-solved.html

the graph could be misinterpreted and should not have been included in these materials."[16]

What about the reference to the workshop white paper that was cited in the IPCC report, in error, as being the source of the graph?

It turns out that Muir-Wood had cited his white paper to our workshop as a placeholder, because he had not yet completed the analysis which he wanted to have displayed in the IPCC report. RMS explained in its press release that a paper based on Muir-Wood's informal analysis was completed after the IPCC's deadline for inclusion in the report, and thus the IPCC had intentionally mis-cited the graph to the workshop paper to get around the IPCC's publication deadline for inclusion of material in the report. RMS wrote in its press release: "Despite not being able to reference it, the IPCC was aware of the full report."[17]

There is more. When that "full report" was eventually published as a book chapter in 2008, the graph did not appear there either. Instead, that chapter concluded, "We find insufficient evidence to claim a statistical relationship between global temperature increase and normalized catastrophe losses."[18] The paper which the IPCC wanted to

---

[16] R. Pielke, Jr., "RMS on the 'Mystery Graph': Should Not Have Been Included," Roger Pielke Jr.'s Blog (17 Feb. 2010), available at: http://rogerpielkejr.blogspot.com/2010/02/rms-on-mystery-graph-should-not-have.html

[17] R. Pielke, Jr., "RMS Confirms Effort to Skirt IPCC Publication Deadlines," Roger Pielke Jr.'s Blog (14 Feb. 2010), available at: http://rogerpielkejr.blogspot.com/2010/02/rms-confirms-effort-to-skirt-ipcc.html

[18] S. Miller, R. Muir-Wood, and A. Boissonnade, "An exploration of trends in normalized weather-related catastrophe losses,"in H. F. Diaz and R. J. Murnane, eds., *Climate Extremes and Society*, (New York, NY: Cambridge University Press, 2008): pp. 225–347.

but couldn't cite to suggest that increasing temperatures were causing increasing damage from extremes actually did not find evidence to support that claim, once it was published.

The IPCC, despite its vaunted peer-review process, had published a claim with no scientific support — either before or since.[19]

For my efforts to hold the IPCC to its own standards of scientific integrity, in the days after our London debate I found myself approached by numerous journalists to comment on issues related to the IPCC. Among them was a reporter named Christina Larson from *Foreign Policy* magazine who sent me an email explaining that *FP* was doing "a sort of guide to what readers should know about the various concerns/criticisms of climate science recently in the news." She asked if she could ask me a few questions about my work. As I typically did with media queries I agreed and answered what seemed to be reasonable, fair questions.

Much to my surprise, the piece published by *Foreign Policy* was not a guide to "concerns/criticisms" but instead a tabloid-like list of the world's "top climate skeptics." They had placed me alongside people like Senator James Inhofe and Christopher Monckton. Larson and the *FP* explained in their piece what had qualified me for the list: "for his work questioning certain graphs presented in the IPCC reports, Pielke has been accused by some of being a climate change 'denier.'"[20]

---

[19] The flawed graph was actually identified in the IPCC review process, and a request was made for it to be removed as "misleading." The IPCC ignored this advice. This episode is chronicled in *The Climate Fix*.

[20] C. Larson and J. Keating, "The FP Guide to Climate Skeptics," *Foreign Policy* (26 Feb. 2010), available at:

For me it was a lesson in the dirty politics of the climate debate. As social commentator and *New York Times* columnist David Brooks has explained, political discourse "is not really a debate about issues; it is a verbal contest to deny your opponents of standing, or as we would say, legitimacy."[21] Who knew that the path from the Roger Revelle lecturer of the National Academy of Sciences to accused climate denier was so short?

The climate kitchen was heating up, but I wasn't going anywhere.

The IPCC Gets Back on Course

In response to various concerns about the integrity of the IPCC and the fidelity of its processes, the United Nations and IPCC asked the InterAcademy Council, comprising the world's leading national science academies, to conduct a review of the organization.[22] The resulting report, published in 2010, recommended a large number of steps to the IPCC to improve its management, treatment of uncertainties, and assessment process. While the IPCC did not accept or implement all of the IAC recommendations, it did take on board many of them.

---

http://www.foreignpolicy.com/articles/2010/02/25/the_fp_guide_to_climate_skeptics

[21] D. Brooks, "The Refiner's Fire," *New York Times* (13 Feb. 2014), available at: http://www.nytimes.com/2014/02/14/opinion/brooks-the-refiners-fire.html

[22] Committee to Review the Intergovernmental Panel on Climate Change, *Climate change assessments: Review of the processes and procedures of the IPCC* (Amsterdam, The Netherlands: InterAcademy Council, 2010), available at:
http://reviewipcc.interacademycouncil.net/report.html

The result of these reforms in how the IPCC reported on the science of extreme weather and disasters became clear in a series of reports issued in 2012, 2013, and 2014.

- In 2012 the IPCC issued a report titled *Managing the Risks of Extreme Events and Disasters to Advance Climate Change Adaptation* (SREX).[23] This report included a focus specifically on disasters and climate change.
- In 2013 the IPCC issued the first volume of its Fifth Assessment Report focused on *The Physical Science Basis* for climate change.[24] This report included a summary of literature related to the frequency and intensity of specific types of extreme events.
- In 2014 the IPCC issued the second volume of its Fifth Assessment Report focused on *Impacts, Adaptation, and Vulnerability*.[25] This report largely reviewed the results of the recent SREX report, but added a few additional details.

One important function of the IPCC is to provide an assessment of the scientific literature, which might contain hundreds or more papers on a particular topic. Its assessment, when the organization is at its best, helps us to un-

---

[23] IPCC, *Managing the Risks of Extreme Events and Disasters to Advance Climate Change Adaptation* (New York, NY: Cambridge University Press, 2012), available at: http://www.ipcc-wg2.gov/SREX/

[24] IPCC, *Climate Change 2013: The Physical Science Basis*, Working Group I Contribution to the Fifth Assessment Report of the Intergovernmental Panel on Climate Change (New York, NY: Cambridge University Press, 2013), available at: http://www.ipcc.ch/report/ar5/wg1/

[25] IPCC, *Climate Change 2014: Impacts, Adaptation, and Vulnerability*, IPCC Working Group II Contribution to AR5 (New York, NY: Cambridge University Press, 2014), available at: http://ipcc-wg2.gov/AR5/

derstand the balance of evidence across a large number of studies, not all of which come to the exact same conclusions. The IPCC can be used to help identify and to avoid the cherry picking of scientific results by providing a broad context for understanding evolving knowledge.

Unlike the 2007 report, the IPCC in 2012, 2013, and 2014 played it straight on climate change and disasters. There were no "mystery graphs" to be found. Instead of alleging a scientifically unsupportable connection between rising temperatures and disaster losses, in these recent reports the IPCC faithfully represented the academic literature.

Here are a few examples of IPCC conclusions from these reports (and these are just examples — subsequent sections will get into much more detail):

- "Economic growth, including greater concentrations of people and wealth in periled areas and rising insurance penetration, is the most important driver of increasing losses."
- "Loss trends have not been conclusively attributed to anthropogenic climate change."
- "Most studies of long-term disaster loss records attribute these increases in losses to increasing exposure of people and assets in at-risk areas (Miller et al., 2008; Bouwer, 2011), and to underlying societal trends — demographic, economic, political, and social — that shape vulnerability to impacts (Pielke Jr. et al., 2005; Bouwer et al., 2007)." [26]
- "Some authors suggest that a (natural or anthropogenic) climate change signal can be found in the records of disaster losses (e.g., Mills, 2005; Höppe and Grimm, 2009), but their work is in the nature of

---

[26] Careful readers will note that in this instance, Miller at al., 2008 is properly cited, unlike in 2007.

reviews and commentary rather than empirical research."

- "There is medium evidence and high agreement that long-term trends in normalized losses have not been attributed to natural or anthropogenic climate change."

Given the strength of these findings, one might think that the issue of climate change and disasters would have become less politicized. However, political battles are often impervious to information, no matter where it originates. It is a lesson I have well understood, but I often have the opportunity for refresher courses.

A "Furious Campaign"

In early 2014 I agreed to join up with Nate Silver's rebranded *FiveThirtyEight*. Silver had moved to ESPN from the *New York Times* following his impressive success tracking polls and integrating them into forecasts of the 2012 U.S. presidential election. I agreed to join up with Silver and his excellent staff in order to write about a range of topics, including sports governance, a topic that has come to occupy an increasing share of my research and writing interests.

However, after discussing options for what to lead off with as the site rolled out, the editors and I decided that my first piece for *FiveThirtyEight* would be on disasters and climate change, summarizing the key findings of the recently-released IPCC reports.[27] There was nothing in that first piece that I had not written on before, including in the peer reviewed literature. Nonetheless, the reaction

---

[27] R. Pielke, Jr., "Disasters Cost More Than Ever — But Not Because of Climate Change," *FiveThirtyEight* (19 Mar. 2014), available at: http://fivethirtyeight.com/features/disasters-cost-more-than-ever-but-not-because-of-climate-change/

to it was remarkable, and made the 2010 sneak attack on me by *Foreign Policy* look like the minor leagues of the delegitimization wars.

The online magazine *Salon* explained that I was "the target of a furious campaign of criticism from other journalists in the field, many of whom say he presents data in a manipulative and misleading way."[28] *Slate* called for me to be fired, and labeled me a "climate change denialist."[29] Paul Krugman, a Nobel Prize-winning economist and *New York Times* columnist, labeled me a "known irresponsible skeptic."[30] The American Geophysical Union, one of the nation's leading scientific associations, published a blog post recommending that Nate Silver should "find an expert on the subject who has many published papers in the top scientific journals (and there are plenty out there), but instead he chose Roger Pielke."[31]

---

[28] E. Isquith, "Objectively bad: Ezra Klein, Nate Silver, Jonathan Chait and return of the 'view from nowhere,'" *Salon* (12 Apr. 2014), available at:
http://www.salon.com/2014/04/12/objectively_bad_ezra_klei n_nate_silver_jonathan_chait_and_return_of_the_view_from_no where/

[29] D. Auerbach, "Unnatural Disaster," *Slate* (31 Mar. 2014), available at: http://www.slate.com/articles/business/moneybox/ 2014/03/nate_silver_climate_change_denial_it_s_time_to_dump _fivethirtyeight_s_roger.html

[30] P. Krugman, "Tarnished Silver," *New York Times* (23 Mar. 2014), available at: http://krugman.blogs.nytimes.com/2014/ 03/23/tarnished-silver/

[31] D. Satterfield, "Lies, Damned Lies and Statistical Truth," AGU Blogosphere website (29 Mar. 2014), available at:
http://blogs.agu.org/wildwildscience/2014/03/29/lies-damned-lies-and-statistical-truth/; the reader can judge whether the suggestion by the AGU that I have not published in top scientific journals is supported by the evidence:
http://scholar.google.com/citations?user=WtqpmdIAAAAJ

These critics were creating their own reality in order to engage in outright character assassination. On *The Daily Show*, Jon Stewart discussed the campaign with Silver and observed, "You are taking a rash of shit in a week and a half like no one I've seen in a long time."[32] That about sums up the tenor and substance of the organized campaign.

The campaign to have me ousted as a writer at *FiveThirtyEight* was nonetheless successful. After subsequently publishing a few of my pieces on sports, Silver eventually refused to publish anything further and we parted ways soon after. I harbor no hard feelings, as Silver was put under an impressive amount of pressure in a setting where being popular seems to be more important than being right. I doubt that Silver will go near the climate and disasters issue again. If you can't take the heat, it is best to get out of the kitchen.

Recalling Harry Truman's sage advice, I am interpreting the furious attacks against me as one reason to prepare this primer on disasters and climate change. Here I'm also throwing in my lot with John Kay, a columnist for the *Financial Times*, who explained in another context about campaigners who try to create their own reality: "Whatever initial misconceptions spin doctors may promote, reality will out."[33]

---

[32] D. Byers, "Why is Nate Silver so sensitive?" *Politico* (28 Mar. 2014), available at: http://www.politico.com/blogs/media/2014/03/why-is-nate-silver-so-sensitive-185900.html

[33] J. Kay, "The welfare cap replaces political judgment with spin," *Financial Times* (1 Apr. 2014), available at: http://www.ft.com/intl/cms/s/0/388faf40-b8fa-11e3-98c5-00144feabdc0.html

2

THE SCIENTIFIC QUESTION ADDRESSED HERE

This short volume addresses a very specific scientific question—a question that can be addressed empirically, that is, with data:

Have disasters become more costly because of human-caused climate change?

There are many other questions which could be asked about climate change, many of which are arguably far more important, but are not addressed here in any significant depth. I have addressed some of these other questions in other venues.[1] However, the central question addressed here stands on its own. It is not a stalking horse for anything else.

Over the years, I have addressed in my work many other questions related to climate change science and policy. Here in a capsule format are some of the other important questions, and my views on them.

Is climate change real?

Yes.

---

[1] See my publications:
http://scholar.google.com/citations?user=WtqpmdIAAAAJ

Does climate change have human causes, notably from the emission of greenhouse gases?

Yes.

Does human-caused climate change pose risks, perhaps significant ones, for life on Earth?

Yes.

Should policy makers around the world take action to reduce emissions toward eventual stabilization of greenhouse gas concentrations?

Yes.

Does a price on carbon make sense?

Yes.

Do scientific projections suggest that some extreme events may become more common or intense?

Yes.

Does current science suggest that episodes of extreme heat and intense rainfall may be increasing in some areas as a consequence of increasing concentrations of greenhouse gases in the atmosphere?

Yes.

Does any of the work summarized in this short volume counter my answers to any of the above questions?

No.

For those who may be interested in my more fully developed perspectives on these questions, I recommend my book, *The Climate Fix* (Basic Books, 2011), which discusses them in detail. I now return to the main question that is the focus of this volume.

Looking for Signals in Complex Data

Studying the relationship between climate change and disasters is challenging because there are many moving parts that contribute to the outcomes that we care about, like property damage or casualties. For instance, the frequency and intensity of extreme events can change and vary over time, but so too does the exposure and vulnerability of human settlements which are subject to experiencing disasters. Much of the work that I have been involved in over the past few decades involves efforts to separate out the role of physical factors (the "natural" in "natural disasters") from societal factors in long-term trends in disaster losses.

One way that researchers make this separation is by adjusting historical loss data to account for relevant societal changes. We seek to standardize losses to a common base year—a process which we have called "normalization."[2] If this procedure is done properly, then trends in the resulting normalized time series should match up well with climatological trends in the relevant extreme events.[3]

For example, according to data kept by the National Hurricane Center, in 2005 Hurricane Katrina caused $80 billion in damage (in what's known as current, or 2005, as opposed to inflation-adjusted, dollars) and in 2012 Superstorm Sandy caused $50 billion (in 2012 dollars). The scale of these numbers makes some sense to us.

But consider the Great Miami Hurricane of 1926. It caused $76 million (in 1926 dollars) of damage almost nine

---

[2] We first used the term "normalization" in this paper: R.A. Pielke, Jr. and C.W. Landsea, "Normalized hurricane damages in the United States: 1925-95," *Weather and Forecasting* 13 (1998): pp. 621-631.

[3] We have applied normalization techniques to earthquakes as well; however, the focus here is on weather and climate.

decades ago, or about 1% of the damage caused by Katrina. That number makes little sense in today's context, because that same storm hitting downtown Miami would surely cause much more damage than either Katrina or Sandy. But exactly how much damage would it cause today?

The answer to this question has great relevance to insurers and reinsurers (those companies that provide insurance to insurance companies), policy makers, and residents of hurricane-prone regions. An entire area of financial analysis called "catastrophe modeling" has developed in the past several decades to address questions like these.

Normalizing disaster losses leads to an estimate that the Great Miami Hurricane of 1926 would cause almost $200 billion in damage were it to hit in 2014. This would make it the most costly hurricane since 1900, if we were to rank all past storms based on what damage each would cause if they hit with today's level of population and development.

The logic here is simple, and can be illustrated with an example. Imagine a house on the beach in 2005. A hurricane comes through and badly damages the house, causing $100,000 in damage. Now imagine that same stretch of the beach ten years later, in 2015. Now, due to coastal development there are two identical houses on the beach. A hurricane of the exact same strength as the earlier storm blows through and damages both houses. The most recent storm causes $200,000 in damage.[4]

In this hypothetical example, storm damage has doubled over a decade. However, the increased damage was not because of stronger or more frequent storms. The in-

---

[4] This simplified example ignores inflationary and wealth effects, which are of course important to address in actual analyses.

crease in damage was entirely due to the doubling of the amount of exposed property.

We would be able to recognize the reason for the increased losses by looking at the data, and asking how much damage the 2005 storm would have caused in 2015. In this simple example, we would simply multiply the 2005 losses by two to arrive at our answer, as there is twice as much exposed property. The 2005 storm, had it occurred in 2015, would have caused $200,000 in damage. Because we have assumed identical storms in this simple example, the only variable that changes is the exposure to damage.

In the real world, however, things are not so simple. Most obviously, houses are built using different practices. So let's consider an alternative scenario. In this second scenario the second home is built with greater attention to damage potential, perhaps with reinforcements or a change in style. When the second storm passes through in 2015 there is only $50,000 in damage, for a total of $150,000 (that is $100,000 damage to the older home, plus $50,000 damage to the newer, stronger home).[5]

Under this second scenario, how much damage would the 2005 storm have caused if it occurred in 2015? If we were to simply multiply the 2005 damage times two — reflecting that in 2015 there are now two houses — we would get $200,000, which is much higher than the $150,000 that is actually observed. We would thus have a bias in our results because we failed to account for the

---

[5] It would be a mistake, however, to assume that building practices inevitably improve over time with respect to loss potentials. In fact, evidence suggests that older homes often do better in storms. See, e.g., B. Tansel and B. Sizirici, "Significance of Historical Hurricane Activity on Structural Damage Profile and Posthurricane Population Fluctuation in South Florida Urban Areas," *Natural Hazards Review* 12 (2011): pp. 196–201.

stronger house of 2015. It would be erroneous to claim from these data that a weaker hurricane occurred in 2015 because of the lesser damage.

We could identify a bias in our results by comparing the normalized losses to the physical characteristics of the storms. Because the storms in this thought experiment are assumed to be identical, we might initially expect damages in 2015 to be twice those of 2005, as we did under the first scenario. The fact that the data do not match up (100% increase in storms but only 50% increase in damage) indicates that there is something left out of our normalization methodology. By comparing trends in damage to trends in storm frequency and intensity, we can check for evidence of a bias in our adjustments.

Now consider a third variation on the thought experiment. In this version, imagine that there are two houses on the beach in 2015, both identical to the single house present in 2005. In this case, a stronger storm makes landfall in 2015, causing $250,000 in damage to the two houses. In this case, after adjusting the 2005 storm to 2015 values, we would see that the normalized damage from the earlier storm had increased from $200,000 to $250,000, and that this increase in damage would be attributable to an increase in storm strength. We would know this not because of the adjusted loss data, but because of the data on storm strength.

Of course our coasts have trillions of dollars in property across many millions of structures. Hundreds of storms over many decades create a complex record of damage and costs. Performing a normalization properly requires paying careful attention to the many societal factors which influence losses, but also to trends in the frequency and intensity of storms.

It seems obvious, but is often overlooked, that in order for climate change, human-caused or otherwise, to contribute to increasing disaster losses, extreme events must become more frequent, more intense, or both. With more frequent or intense events, we would expect to see similar increases in normalized losses. Because data on the frequency and intensity of extreme weather events are usually collected separately from dollar losses, these two independent types of data provide a very important consistency check for any normalization process. Trends in each dataset should match up. If they don't then there remains something to be explained in the data.

When we conduct research to adjust past loss events to present day values, we are in effect asking how much damage would occur today if events of the past occurred with today's level of population and economic development.

So the first step in evaluating any normalization of disaster losses is to see if the trends in the adjusted losses correspond with trends in the frequency or intensity of the relevant events. If they don't then there is a remaining bias in the procedure which needs to be addressed. You will see some numbers on this type of check in the sections which follow, but the logic here is simple: Trends in normalized losses should match up with trends in the relevant weather events.

It is important to underscore that we do not look at normalized loss data in order to identify changes in extreme events or to discover a "signal" of climate change. The best place to look for evidence of changes in the frequency or intensity of extreme events is, not surprisingly, in data which directly reflect those extreme events.[6] Scholars are well aware of this issue, even if it does not make it into popular discussions of disaster losses, where

---

[6] L.M. Bouwer, R.P. Crompton, E. Faust, P. Höppe, and R.A. Pielke, Jr., "Confronting disaster losses," *Science* 318 (2007): p. 753ff.

growing damage by itself is sometimes simply assumed to reflect climate trends. At the conclusion of the next chapter, I summarize in a flow chart the role of normalization research in the search for a signal of human-caused climate change in the growing toll of disasters.

As you will see in the chapters which follow, disaster losses have increased dramatically over recent decades. However, once past losses are adjusted for societal changes (more houses, more possessions, and so on), there is no remaining increase, leaving essentially no residual trend to be explained. In the several dozen normalization studies that have been published for phenomena around the world, you will find many different approaches to normalization. More frequent or intense extreme events, whatever the cause, are not necessary to explain the dramatic increase in disaster losses. Not only are they not necessary, but the evidence on extreme events is perfectly consistent with the normalization results.

But I am getting ahead of myself. We will get to the data shortly, but first it is necessary to address a common complaint.

Can Science Prove a Negative?

Over the many years that I have worked on disasters and climate change, a common response to my work and that of my colleagues has been that it does not conclusively prove that human-cause climate change is *not* influencing extreme events.

Not only are such responses absolutely true, but they are in fact truisms. Science cannot prove a negative.

All that we can say is that the record of disaster losses is entirely explainable by changes in society. There is at present no evidence that human-caused climate change is responsible for any part of the global increase in disaster costs. We cannot say that there is no such influence.

But as I have explained on many occasions, from a practical standpoint a signal that may exist, but which cannot be detected, is indistinguishable from a signal that does not exist.

Yet the desire to prove a negative persists.[7] Some borrow a phrase most commonly associated with arguments over the existence of God and aliens: "absence of evidence is not evidence of absence."[8] This catchy phrase is of course a tautology, hardly helpful and of little relevance to the central question of this volume. Science is concerned with evidence, not with supporting pre-existing beliefs.

The philosopher Bertrand Russell provided a useful analogy for such arguments:

---

[7] Even the IPCC has adopted the framing of proving a negative in at least one place. Chapter 18 of AR5 WGII concludes: "climate change cannot be excluded as at least one of the drivers involved in changes of normalized losses over time in some regions and for some hazards." This statement is a truism, and says absolutely nothing of substance. Science does not prove negatives. See "Detection and Attribution of Observed Impacts," Ch. 18 in IPCC, *Climate Change 2014: Impacts, Adaptation, and Vulnerability*, IPCC Working Group II Contribution to AR5 (New York, NY: Cambridge University Press, 2014), available at: http://ipcc-wg2.gov/AR5/images/uploads/WGIIAR5-Chap18_FGDall.pdf

[8] E. Morris, "The Certainty of Donald Rumsfeld (Part 4)," *New York Times* (28 Mar. 2014), available at: http://opinionator.blogs.nytimes.com/2014/03/28/the-certainty-of-donald-rumsfeld-part-4/; for an example of its use in a climate context, see A. Revkin, "Varied Views on Extreme Weather in a Warming Climate," *New York Times* (11 May 2012), available at: http://dotearth.blogs.nytimes.com/2012/05/11/another-view-on-extreme-weather-in-a-warming-climate/

Many orthodox people speak as though it were the business of skeptics to disprove received dogmas rather than of dogmatists to prove them. This is, of course, a mistake. If I were to suggest that between the Earth and Mars there is a china teapot revolving about the sun in an elliptical orbit, nobody would be able to disprove my assertion provided I were careful to add that the teapot is too small to be revealed even by our most powerful telescopes. But if I were to go on to say that, since my assertion cannot be disproved, it is an intolerable presumption on the part of human reason to doubt it, I should rightly be thought to be talking nonsense.[9]

Russell continues, explaining that if the belief in the celestial teapot were widely affirmed and instilled, then "hesitation to believe in its existence would become a mark of eccentricity and entitle the doubter to the attentions of the psychiatrist in an enlightened age or of the Inquisitor in an earlier time." While Russell was referring to arguments over the existence of God, he might as well have been talking about the climate debate circa 2014.

The state of climate science today suggests that we should no more expect to see a signal of human-caused climate change in the increasing disaster losses of the past several decades than we should expect to find a teapot orbiting the sun.[10] More precisely, the IPCC's climate

---

[9] B. Russell, *Is There a God? The Collected Papers of Bertrand Russell, Vol. 11: Last Philosophical Testament, 1943-68* (London, UK: Routledge, 1952): pp. 547–548.

[10] The IPCC discusses projections for near-term climate here: "Near-term Climate Change: Projections and Predictability," Ch. 11 (http://www.climatechange2013.org/images/report/WG1AR5_Chapter11_FINAL.pdf); and longer-term climate here: "Long-term Climate Change: Projections, Commitments and Irreversibility, Ch. 12 (http://www.climatechange2013.org/images/report/WG1AR5_Chapter12_FINAL.pdf) in IPCC, *Climate Change 2013: The Physical Science Basis*, Working Group I Contribution to the Fifth Assessment Report of the Intergovern-

model projections of changes in extreme events do not show identifiable increases in disaster losses for many decades, and often much longer.[11]

Richard Dawkins, a long-time partisan in debates over religion and science, argues that a focus on the significance of an "absence of evidence" encourages "sloppy thinking":

> *Agnostic conciliation, which is the decent liberal bending over backward to concede as much as possible to anybody who shouts loud enough, reaches ludicrous lengths in the following common piece of sloppy thinking. It goes roughly like this: You can't prove a negative (so far so good). Science has no way to disprove the existence of a supreme being (this is strictly true). Therefore, belief or disbelief in a supreme being is a matter of pure, individual inclination, and both are therefore equally deserving of respectful attention! When you say it like that, the fallacy is almost self-evident; we hardly need spell out the reductio ad absurdum.[12]*

It is of course true that a role for climate change in the growing toll of disaster losses has not been excluded in any of the studies or assessments that are discussed here. In fact, such exclusion is a logical impossibility. Science cannot by its nature prove a negative. It would be equally true to state that science does not exclude a role for solar

---

mental Panel on Climate Change (New York, NY: Cambridge University Press, 2013).

[11] For the basic mathematics of such a calculus expressed in the case of U.S. hurricanes see, e.g., K. Emanuel, "Global Warming Effects on U.S. Hurricane Damage," *Weather, Climate, and Society* 3, vol. 4 (2011): pp. 261-268; and R.P. Crompton, R.A. Pielke, Jr., and K.J. McAneney, "Emergence timescales for detection of anthropogenic climate change in U.S. tropical cyclone loss data," *Environmental Research Letters* 6, vol. 1 (2011): 014003.

[12] R. Dawkins, "Snake Oil and holy Water," *Forbes* (4 Oct. 1999), available at: http://www.forbes.com/asap/1999/1004/235_2.html

influences, cosmic rays, or, for that matter, evil lepre-chauns in explaining trends in disaster losses.

The good news about the subject of disasters and climate change is that there is lots of evidence to look at; we don't need to rely on clever logical constructions. The following sections of this volume assess some of the evidence with respect to the central question:

Have disasters become more costly because of human-caused climate change?

There are two possible answers to this question:

1. Yes, the evidence indicates that disasters have become more costly because of climate change.
2. No, there is not sufficient evidence to indicate that disasters have become more costly because of climate change.

There is a third possible position of course — agnosticism. However, given the strength and depth of the research in this area, the only legitimate reason for agnosticism is a lack of awareness of the relevant science and data. If you keep reading this volume, you won't be able to take that position!

A good analogy here would be evidence in support of claims that the Earth has warmed, on average, over the past century or so. The strength and depth of research on this topic indicates that it has. But it does not rule out the possibility that it has not warmed, of course (remember, science cannot prove a negative). The IPCC reports its findings with degrees of certainty, which never reach 100%. The degree of certainty expressed by the IPCC with respect to warming is similar to its degree of certainty about disasters and climate change.

As a tonic against lack-of-awareness-induced agnosticism, the next sections present some of the data that help us to address the central question which is the focus of this volume.

Have a look, and come to your own conclusions about what the evidence says.

3

THE IPCC FRAMEWORK FOR
DETECTION AND ATTRIBUTION

This section describes the framework of detection and attribution which underlies the work of the IPCC, and how it has been applied in studies that focus on the central question of this book.

I have noticed in my years writing and speaking on climate change that what "climate change" actually means is not widely understood in a common way. For instance, the IPCC and the UN Framework Convention on Climate Change (and its most widely known policy agreement, the Kyoto Protocol) use different definitions of "climate change."[1]

Here I explain in some detail what it means to detect and attribute a signal of human-caused climate change in the disaster record, so that there is no confusion by what I mean when I discuss this topic. I rely on the IPCC definitions here, not because they are necessarily the best or final words on these subjects, but rather because they are

---

[1] R.A. Pielke, Jr., "Misdefining 'climate change': consequences for science and action," *Environmental Science and Policy* 8 (2005): pp. 548-561.

widely accepted and used in the climate science community.

Let's start with the IPCC's definition of "climate":

Climate in a narrow sense is usually defined as the average weather, or more rigorously, as the statistical description in terms of the mean and variability of relevant quantities over a period of time ranging from months to thousands or millions of years.[2]

"Weather" refers to "the conditions of the atmosphere at a certain place and time with reference to temperature, pressure, humidity, wind, and other key parameters (meteorological elements)." Extreme weather events include phenomena such as heat waves, winter storms, tropical cyclones, floods, and so on. Weather events occur over minutes, hours, days, and perhaps even weeks.

The IPCC also defines what it means by "climate change":

Climate change refers to a change in the state of the climate that can be identified (e.g., by using statistical tests) by changes in the mean and/or the variability of its properties, and that persists for an extended period, typically decades or longer.

It is important to understand that the IPCC definition of "climate change" makes no reference to the cause of the observed change. It is simply an identifiable change in the statistical properties of climate over a fairly long time period, which the IPCC identifies as "decades or longer." Because many extreme events are—by definition—rare

---

[2] "Introduction," Ch. 1 in IPCC, *Climate Change 2013: The Physical Science Basis*, Working Group I Contribution to the Fifth Assessment Report of the Intergovernmental Panel on Climate Change (New York, NY: Cambridge University Press, 2013), available at: http://www.climatechange2013.org/images/report/WG1AR5_Chapter01_FINAL.pdf

events, the time scale for the detection of change will necessarily be longer than that for variables which are measured more frequently, such as daily weather, precipitation, or sea level rise.

Even in the absence of detectable "climate change," the occurrence of weather events varies on all time scales from seconds to millennia. The IPCC calls this "climate variability":

Climate variability refers to variations in the mean state and other statistics (such as standard deviations, the occurrence of extremes, etc.) of the climate at all spatial and temporal scales beyond that of individual weather events.[3]

The presence of climate variability is one of the most significant obstacles that scientists must overcome in detecting a change in climate. For instance, there is a broad consensus that there has been greater hurricane activity in the North Atlantic since 1970. However, there is also a broad consensus that the increase since 1970 falls within the variability observed in North Atlantic hurricanes observed since 1900.[4] Thus, "climate change" as defined by the IPCC has not been detected with respect to hurricanes.

---

[3] "Glossary of Terms," Annex II in IPCC, *Managing the Risks of Extreme Events and Disasters to Advance Climate Change Adaptation* (New York, NY: Cambridge University Press, 2012), available at: http://www.ipcc.ch/pdf/special-reports/srex/SREX-Annex_Glossary.pdf

[4] The IPCC concludes: "No robust trends in annual numbers of tropical storms, hurricanes and major hurricane counts have been identified over the past 100 years in the North Atlantic basin." See: "Observations: Atmosphere and Surface," Ch. 2 in IPCC, *Climate Change 2013: The Physical Science Basis*, Working Group I Contribution to the Fifth Assessment Report of the Intergovernmental Panel on Climate Change (New York, NY: Cambridge University Press, 2013), available at: http://www.climatechange2013.org/images/report/WG1AR5_Chapter02_FINAL.pdf

It is also important to understand that no discernible change in particular extreme weather events does not mean that they or other climate metrics are not changing. A change may be underway, but will not be detectible until sometime in the future. For example, as noted above, under recent model projections assuming that greenhouse gas emissions will influence North Atlantic hurricane behavior, a change in the statistics of hurricanes is expected under some model projections, but it would not be detectable for many decades. The magnitude of the ongoing changes is simply too small in the context of existing variability to be detected in the near future.

Once a change in climate is detected, scientists then ask another difficult question: why has the observed change occurred?

The answering of this question is called "attribution":

Attribution of causes of climate change is the process of establishing the most likely causes for the detected change with some defined level of confidence.[5]

For example, with respect to North Atlantic hurricanes, the fact that there has not been detection of a change in the statistics of storms since 1900 means that there is not a climate change signal to be attributed. This stands in contrast to the robust detection of an increase in global average surface temperatures since the 19th century, which the IPCC attributes with high levels of certainty to human causes.

Even within the IPCC definitions there is considerable room for debate and different perspectives. For instance,

---

[5] "Glossary of Terms," Annex II in IPCC, *Managing the Risks of Extreme Events and Disasters to Advance Climate Change Adaptation* (New York, NY: Cambridge University Press, 2012), available at: http://www.ipcc.ch/pdf/special-reports/srex/SREX-Annex_Glossary.pdf

Kerry Emanuel, a climate scientist at the Massachusetts Institute of Technology, recently discussed the challenges of understanding hurricane behavior in the North Atlantic since 1970:

In the Atlantic, demonstrably hurricane power has increased over the last 30 years by a big factor, too. I don't profess to understand that. It's gone up hand in hand with the tropical Atlantic surface temperature in the summer time. It's a tiny piece of the globe. And maybe some of that is global warming. I don't honestly know. I don't want to try to give you the illusion that I understand this.[6]

In the public discussion of climate change there are often two types of confusion that show up.

One is confusion between climate variability and climate change. For instance, by some measures global average surface temperature has slowed its rate of increase or even paused over the past decade or so. Some point to this as evidence that climate change, as measured by the warming of global average surface temperature, has stopped. While such a slowdown might give scientists good reasons to ask hard questions of climate model projections, it has been going on for too short a time period to understand what, if anything, it might be telling us about changes in climate, which is only discernible on longer time scales. The IPCC explains that climate variability "diminishes the relevance of trends over periods as short as 10–15 years for long-term climate change."[7]

---

[6] "John Christy and Kerry Emanuel on Climate Change," *EconTalk Podcast*, Library of Economics and Liberty (24 Mar. 2014), available at: http://www.econtalk.org/archives/2014/03/john_christy_an.html

[7] "Evolution of Climate Models," Ch. 9 in IPCC, *Climate Change 2013: The Physical Science Basis*, Working Group I Contribution to the Fifth Assessment Report of the Intergovernmental Panel on Climate Change (New York, NY: Cambridge University Press,

With respect to extreme events, such confusion is common, even down to the level of individual weather events or seasons, which are often cited in isolation as evidence for or against the role of human influences on the climate system. For instance, extreme winters are pointed to by some as evidence counter to theories of human-caused climate change, while individual hurricanes or tornadoes are used as evidence of human-caused climate change. Of course, all of this is imbued with a heavy overlay of politics. The fact is that the shorter the time period, the less relevant it is to understanding longer-term climate change and its causes.

A second common confusion is to conflate detection and attribution. Observing a change in a weather variable is not the same thing as associating that change with a particular cause, such as changes in climate resulting from greenhouse gas emissions. In popular discourse this distinction is frequently lost, with trends in any variable automatically assumed to be caused by emissions of greenhouse gases. This is most clearly represented in the common usage of the phrase "climate change" as synonymous with "human-caused climate change," despite the IPCC's broader definition.

This second type of confusion can further be seen when the phrase "climate change" is used itself as a causal factor. For example, whenever there is an extreme event there are inevitably many media stories that ask "did climate change cause X?"[8]

This question is inherently nonsensical. "Climate change" is not a causal actor. It is a statistical property that reflects the consequences of causes.

2013), available at: http://www.climatechange2013.org/images/uploads/WGIAR5_WGI-12Doc2b_FinalDraft_Chapter09.pdf

[8] A Google search of "did climate change cause" returns more than 100,000 results.

Imagine a baseball player who steps up to the plate and knocks a pitch out of the park. Home run! It so happens that this season his batting average is an impressive 0.320, after a sub-par performance of 0.220 last year.

Would we say, "The home run was caused by batting-average change"? Of course not. That would be circular and empty.

His batting average is a measure of change in his hitting. That measure is not a reason why he hits better, but a description of that change. Maybe he practiced more, had laser surgery on his eyes, or is taking performance-enhancing drugs. Or maybe he is just lucky. These might all be causal explanations for his improved hitting. "Batting average change" is not.

Unfortunately, much discussion of climate change is also circular and empty in exactly the same manner. "Climate change" no more causes weather events than changes in batting averages cause home runs.

In the climate debate both types of confusion often occur simultaneously, with extreme events routinely associated with greenhouse gas emissions regardless of whether or not detection or attribution has been achieved. Sometimes this reflects clever political expediency. Sometimes it just reflects confusion. In either case, it is simply wrong.

The IPCC definitions and its framework for detection and attribution overviewed here provide a logical approach to addressing the central question of this short volume:

Have disasters become more costly because of human-caused climate change?

In order for research to show that disasters have become more costly because of human-caused climate change, several criteria must be met.

There must be a *detectable increase* in either the frequency or intensity of weather events, on climate time scales, which are associated with the disasters.

The detected increase in frequency or intensity must be *attributed to human causes*, typically defined narrowly in terms of greenhouse gas emissions, but other causes are also possible.

The following page shows a flowchart which illustrates the necessary and sufficient conditions for the detection and attribution of a role for human-caused climate change in increasing disaster losses. The flow chart also incorporates the concept of "normalization" which was introduced in the previous chapter.

The next chapters will survey a selected number of studies focused at the global level, and then several case studies focused on individual phenomena, and will consider detection and attribution.

Figure 3.1: The Steps Necessary and Sufficient to Achieve Detection and Attribution, under the IPCC Framework, of a Role for Human-Caused Climate Change

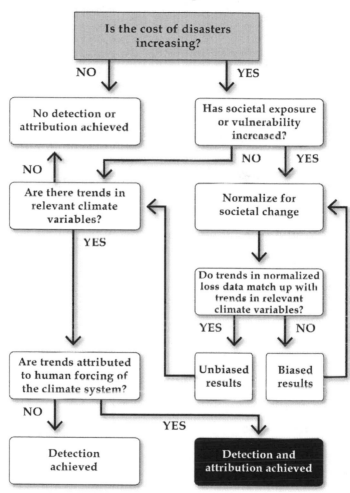

4

A GLOBAL PERSPECTIVE ON DISASTERS AND CLIMATE CHANGE

The IPCC offers a useful definition of a disaster:

Severe alterations in the normal functioning of a community or a society due to hazardous physical events interacting with vulnerable social conditions, leading to widespread adverse human, material, economic, or environmental effects that require immediate emergency response to satisfy critical human needs and that may require external support for recovery.[1]

For present purposes, there are two parts of this definition that are important to highlight.

First, a disaster results from the intersection of a physical event and a vulnerable society. This implies that to understand changes in weather-related disasters over time, we need to focus on both how physical events may have changed and how societal vulnerability may have changed. Such understandings are not just of academic interest, but they also can help to shape our thinking

---

[1] "Glossary of Terms," Annex II in IPCC, *Managing the Risks of Extreme Events and Disasters to Advance Climate Change Adaptation* (New York, NY: Cambridge University Press, 2012), available at: http://www.ipcc.ch/pdf/special-reports/srex/SREX-Annex_Glossary.pdf

about policy options to better prepare for and respond to future extreme events.

A second important part of the definition is its recognition that the effects of a disaster can be measured in many different dimensions—such as loss of life, property damage, loss of social capital, or environmental impacts. The focus of this volume is on economic losses, which primarily occur through damage to physical property, both public and private, as a consequence of extreme weather events. The reason for this focus is not that other types of losses are unimportant, but because the capacity to measure economic losses with some precision gives us a better chance of determining empirically whether human-caused climate change has made disasters worse, which is the central focus of this volume.

The main extreme weather events which cause property damage are wind storms and floods. According to data kept by Munich Re, from 1980 to 2013 almost 80% of all events causing losses worldwide (as measured above a certain threshold) came from wind storms (tropical cyclones, winter storms, thunderstorms) and floods.[2] If we can understand what is driving increasing losses with respect to wind storms and floods, then we will have gone a long way to understanding what is driving disaster trends overall.

So far, there have been five peer-reviewed studies at the global scale looking to disentangle social and climate factors which may underlie loss trends. Four of these studies examined the Munich Re dataset and the fifth looked at a different dataset. All five reach consistent con-

---

[2] "2013 Natural Catastrophe Year in Review," Munich Re (7 Jan. 2014), available at: https://www.munichre.com/site/touch-publications/get/documents_E2138584162/mr/assetpool .shared/Documents/5_Touch/Natural%20Hazards/NatCatNe ws/2013-natural-catastrophe-year-in-review-en.pdf

clusions, despite using different approaches in their analyses.

The studies are:

(1) A 2006 expert workshop that I helped to organize with Munich Reinsurance;
(2) The 2008 RMS study that was supposed to be the source for the 2007 IPCC "mystery graph";
(3) A 2011 study funded by Munich Re and conducted by scholars at the London School of Economics;
(4) A 2014 study that I collaborated on, which looked at the Munich Re global dataset in the context of studies of other more localized datasets; and
(5) A 2014 study which looked at a Belgian research center's data on disaster losses, focusing on economic losses as well as other metrics.

Let's briefly consider each study in turn.

1. The 2006 Hohenkammer Workshop

In 2006, as a contribution to the work of the IPCC Fourth Assessment Report (AR4), I helped to organize a workshop, sponsored by Munich Re and research bodies in the U.S., United Kingdom, and Germany. The workshop included experts from around the world, and sought to assess the literature and reach consensus on the state of the science on disasters and climate change.

We reached a number of unanimous conclusions, among them:

- Analyses of long-term records of disaster losses indicate that societal change and economic development are the principal factors responsible for the documented increasing losses to date.
- Because of issues related to data quality, the stochastic nature of extreme event impacts, length of

time series, and various societal factors present in the disaster loss record, it is still not possible to determine the portion of the increase in damages that might be attributed to climate change due to greenhouse gas (GHG) emissions.

- In the near future the quantitative link (attribution) of trends in storm and flood losses to climate changes related to GHG emissions is unlikely to be answered unequivocally.[3]

We subsequently published these conclusions in *Science* in 2007.[4] As we saw in Chapter 1, the workshop results were selectively mis-cited by the IPCC AR4 report. In fact, the workshop arrived at conclusions exactly the opposite to the claims advanced by the IPCC.

2. The "Mystery Graph" Study

Another examination of the Munich Re dataset can be found in the denouement to the "mystery graph" episode from the 2007 IPCC AR4 report, which I discussed previously. Robert Muir-Wood and colleagues at RMS looked at the Munich Re data from 1980 to 2005 and reached the following conclusions when their paper was finally published in 2008:

---

[3] P. Höppe and R. Pielke, Jr., eds., *Workshop on Climate Change and Disaster Losses: Understanding and Attributing Trends and Projections*, final workshop report (Hohenkammer, Germany: Munich Re and University of Colorado, 25-26 May 2006), available at: http://cstpr.colorado.edu/sparc/research/projects/extreme _events/munich_workshop/workshop_report.html

[4] L.M. Bouwer, R.P. Crompton, E. Faust, P. Höppe, and R.A. Pielke, Jr., "Confronting disaster losses," *Science* 318, No. 5851 (2007): pp. 753-753.

> [T]he large portion of the rising loss trend is explained by
> increases in values and exposure as well as by an increasing
> comprehensiveness of reporting global losses through time.

With respect to a climate signal in the loss record, they
reached the following conclusions:

> In sum, we found limited statistical evidence of an upward
> trend in normalized losses from 1970 through 2005 and in-
> sufficient evidence to claim a firm link between global warm-
> ing and disaster losses. Our findings are highly sensitive to
> recent U.S. hurricane losses, large China flood losses, and
> interregional wealth differences. When these factors are ac-
> counted for, evidence for an upward trend and the relation-
> ship between losses and temperature weakens or disappears
> entirely.[5]

One interesting aspect of this study was that its final
year of analysis was 2005, which saw Hurricane Katrina
and its massive economic impacts. With the presence of a
big loss year at the end of the dataset, it would enhance
any trend. Even so, the authors find little or no evidence
of a relationship of increasing global temperatures ("global warming") and increasing disaster losses.

3. The Munich Re LSE Project

Soon thereafter Munich Re provided financial support
to the London School of Economics for a large research
project to re-examine the same dataset. The scholars at
LSE applied two methods to their normalization, looking
at trends in losses since 1980.

They concluded in a paper published in 2011:

---

[5] S. Miller, R. Muir-Wood, and A. Boissonnade, "An exploration
of trends in normalized weather-related catastrophe losses," in
H. F. Diaz and R. J. Murnane, eds., *Climate Extremes and Society*
(New York, NY: Cambridge University Press, 2008): pp. 225–347.

Independently of the method used, we find no significant upward trend in normalized disaster loss. This holds true whether we include all disasters or take out the ones unlikely to be affected by a changing climate. It also holds true if we step away from a global analysis and look at specific regions or step away from pooling all disaster types and look at specific types of disaster instead or combine these two sets of dis-aggregated analysis.

Much caution is required in correctly interpreting these findings. What the results tell us is that, based on historical data, there is no evidence so far that climate change has increased the normalized economic loss from natural disasters. More cannot be inferred from the data.[6]

In particular, they caution against using loss data, even after it is normalized, to reach conclusions about how specific types of weather events may or may not be changing. The general lesson, one widely accepted and as discussed above, is that if you want to look for changes in the frequency or intensity of extreme weather phenomena, it is always best to look at data on extreme weather phenomena.

4. Trend Reconciliation

In 2014 Shalini Mohleji of the American Meteorological Society (who had previously been a student of mine) and I published a paper which attempted to close the circle on this research by disaggregating the Munich Re dataset into its component parts, organized by phenomena and region. We then compared the trends in the disaggregated data from the Munich Re data set with independent analyses of losses for specific phenomena in various regions.

---

[6] E. Neumayer and F. Barthel, "Normalizing economic loss from natural disasters: A global analysis," *Global Environmental Change* 21 (2011): pp. 13-24.

In many cases data are available for specific phenomena that go much further back in time than 1980. For instance, disaster loss data on U.S. hurricanes go back to 1900. We wanted to assess the consistency between the Munich Re data and the broader literature.

We concluded:

To sum, at the regional level, analyses of normalized damage over time periods longer than but encompassing the data covered by Munich Re's dataset, show no evidence of an anthropogenic climate change signal in economic loss trends for phenomena which account for 97% of the documented increase in losses 1980-2008.[7]

In the next chapter I'll discuss some of these regional studies for specific phenomena in greater detail.

5. A Normalization of the EM-DAT Dataset

In 2014 researchers in the Netherlands and the UK published a paper which looked at a dataset kept by a Belgian research group. Their results, using a different dataset than that which was the focus of the other four papers reviewed above, were nonetheless very similar to those other studies.

Specifically, once the data was normalized, the researchers concluded:

---

[7] S. Mohleji and R. Pielke, Jr., "Reconciliation of Trends in Global and Regional Economic Losses from Weather Events: 1980–2008," *Natural Hazards Review* 15 (2014), available at: http://dx.doi.org/10.1061/(ASCE)NH.1527-6996.0000141

The absence of trends in normalized disaster burden indicators appears to be largely consistent with the absence of trends in extreme weather events.[8]

The paper concluded that the lack of trends in extreme weather events and normalized losses indicates that, overall, vulnerability to losses has been constant over time. The lack of a detectable change in vulnerability at the global level does not preclude changes in vulnerability at more localized contexts, only that any such signal is not detectable at the global level once data has been normalized to account for changes in exposure to loss.

Conclusion

At the global level, it should be clear that the available evidence provides no support for claims that disaster losses have been increasing due to climate change, whether those changes are human-caused or not. Once societal factors are taken into consideration, there is no residual trend. In the language of the IPCC, detection has not been achieved. There is consequently no remaining increase in losses to be attributed to any factors beyond the various societal factors which lead to increasing disaster losses.

Given these studies, it should not come as a surprise that the IPCC's 2012 special report on extreme events came to exactly this conclusion, which was based on regional studies as well as the global studies reviewed here:

Long-term trends in economic disaster losses adjusted for wealth and population increases have not been attributed to

---

[8] H. Visser, A.C. Petersen, and W. Ligtvoet, "On the relation between weather-related disaster impacts, vulnerability and climate change," *Climatic Change* 125, Nos. 3-4 (2014): pp. 461-477.

climate change, but a role for climate change has not been excluded (medium evidence, high agreement).[9]

Its 2014 report on impacts and vulnerability reinforced these conclusions:

- "Economic growth, including greater concentrations of people and wealth in periled areas and rising insurance penetration, is the most important driver of increasing losses."
- "Apart from detection, loss trends have not been conclusively attributed to anthropogenic climate change; most such claims are not based on scientific attribution methods."[10]

These conclusions are very strong (even though they also continue to fall back on the meaningless assertion that the negative has not been proven). Let's now take a closer look at some of the various regional studies for the U.S. and the world, which will reinforce these conclusions at a more localized level.

---

[9] IPCC, *Managing the Risks of Extreme Events and Disasters to Advance Climate Change Adaptation* (New York, NY: Cambridge University Press, 2012), available at: http://www.ipcc-wg2.gov/SREX/

[10] "Key Economic Sectors and Services," Ch. 10 in IPCC, *Climate Change 2014: Impacts, Adaptation, and Vulnerability*, Working Group II Contribution to the Fifth Assessment Report of the Intergovernmental Panel on Climate Change (New York, NY: Cambridge University Press, 2014), available at: http://ipcc-wg2.gov/AR5/images/uploads/WGIIAR5-Chap10_FGDall.pdf

5

HEAT, RAIN, HURRICANES, FLOODS, TORNADOES, DROUGHT, OH MY!

This chapter surveys, in a rather dry and academic manner, some of the literature on climate change and extreme events with a focus on detection and attribution. I rely heavily on the recent assessments of the IPCC, which has done considerable work to assess the relevant literature. However, those passages relevant to the IPCC's discussion of extreme events are spread over dozens of chapters in multiple reports. This section aims to provide a summary which is perfectly consistent with the findings of the IPCC. Consequently, I reproduce a number of extended excerpts from its reports.

Specifically, this chapter focuses on the following phenomena:

- Extreme heat
- Extreme precipitation
- Tropical cyclones (hurricanes)
- Floods
- Tornadoes
- Drought (including Australian bushfire)

This chapter covers a lot of territory, but only scratches the surface of what is available in the primary literature.

In 2011 Dutch researcher Laurens Bouwer wrote a review paper summarizing much of the literature on disaster losses and climate change available at that time. That review paper concluded:

> *The analysis of twenty-two disaster loss studies shows that economic losses from various weather related natural hazards, such as storms, tropical cyclones, floods, and small-scale weather events such as wildfires and hailstorms, have increased around the globe. The studies show no trends in losses, corrected for changes (increases) in population and capital at risk, that could be attributed to anthropogenic climate change. Therefore it can be concluded that anthropogenic climate change so far has not had a significant impact on losses from natural disasters.* [1]

This conclusion should by now be familiar. But let's now take a closer look at some of the studies of specific phenomena at the regional level. Those wanting to explore further are directed to Bouwer's paper, and to the more recent literature summarized by the IPCC.

In the language of the IPCC, at the global scale detection of increases in extreme heat and extreme precipitation has been achieved. These increases have also been attributed to human causes, specifically to growing concentrations of greenhouse gases in the atmosphere. Of course the IPCC expresses all of its conclusions not with 100% certainty, but accompanied by a judgment call about the confidence they have in their conclusions.

With respect to tropical cyclones, floods, tornadoes, and drought, neither detection nor attribution has been achieved at the global scale. At selected regional and local scales there is evidence of trends over various time

---

[1] L.M. Bouwer, "Have disaster losses increased due to anthropogenic climate change?" *Bulletin of the American Meteorological Society* 92 (2011): pp. 39-46.

periods, which of course would be expected as the Earth's climate system is highly variable. Attribution of small-scale, short-time period trends to anything other than variability of the climate system remains a challenge. Let's now look at some of the IPCC's findings and some of the underlying data for these various phenomena.

Extreme Temperature and Precipitation

Extreme temperatures and precipitation are not big drivers of disaster losses. There are nonetheless important phenomena with significant impacts on people and ecosystems, and they have been changing.

With respect to extreme temperatures, the IPCC Fifth Assessment Report (AR5) concludes:

> [T]here is medium confidence that globally the length and frequency of warm spells, including heat waves, has increased since the middle of the 20th century although it is likely that heatwave frequency has increased during this period in large parts of Europe, Asia and Australia.[2]

In the dry prose of the IPCC "medium confidence" is certainly not as sensational as some characterizations of the IPCC conclusions. For the U.S., the IPCC concluded with respect to heat waves: "*Medium confidence*: increases in more regions than decreases but 1930s [Dust Bowl] dominates longer term trends in the USA."

---

[2] "Observations: Atmosphere and Surface," Ch. 2 in IPCC, *Climate Change 2013: The Physical Science Basis*, Working Group I Contribution to the Fifth Assessment Report of the Intergovernmental Panel on Climate Change (New York, NY: Cambridge University Press, 2013), available at: http://www.climatechange2013.org/images/report/WG1AR5_Chapter02_FINAL.pdf

With respect to attribution, the IPCC surveys a large number of modeling studies which try to disentangle human forcing of the climate system from ongoing climate variability. The AR5 concludes from this research:

> [N]ew results suggest more clearly the role of anthropogenic forcing on temperature extremes compared to results at the time of the SREX assessment. We assess that it is very likely that human influence has contributed to the observed changes in the frequency and intensity of daily temperature extremes on the global scale since the mid-20th century.[3]

In the jargon of the IPCC, "very likely" means that they are expressing a likelihood value of at least 90% probability that this claim is correct.[4]

With respect to extreme precipitation the IPCC's conclusions are not nearly as strong as those for extreme temperatures. The IPCC concludes:

> [I]t is likely that since 1951 there have been statistically significant increases in the number of heavy precipitation events (e.g., above the 95th percentile) in more regions than there have been statistically significant decreases, but

---

[3] "Detection and Attribution of Climate Change: from Global to Regional," Ch. 10 in IPCC, *Climate Change 2013: The Physical Science Basis*, Working Group I Contribution to the Fifth Assessment Report of the Intergovernmental Panel on Climate Change (New York, NY: Cambridge University Press, 2013), available at: http://www.climatechange2013.org/images/report/WG1AR5_Chapter10_FINAL.pdf

[4] IPCC, "Guidance Note for Lead Authors of the IPCC Fifth Assessment Report on Consistent Treatment of Uncertainties," IPCC Cross-Working Group Meeting (6-7 Jul. 2010), available at: http://www.ipcc.ch/pdf/supporting-material/uncertainty-guidance-note.pdf

there are strong regional and subregional variations in the trends.[5]

By "likely" the IPCC means that there is at least a 66% likelihood that this particular claim is correct. The IPCC thus judges that there is a two in three chance that there have been increases in heavy precipitation in more locations than have seen decreases.

Given the large uncertainties in detection of trends, it is therefore not surprising that the IPCC expressed limited confidence in attribution:

[T]here is medium confidence that anthropogenic forcing has contributed to a global scale intensification of heavy precipitation over the second half of the 20th century in land regions where observational coverage is sufficient for assessment.[6]

The IPCC expressed "high confidence" in its conclusions that precipitation extremes had "very likely" increased in central North America, its strongest conclusion for any region.

---

[5] "Observations: Atmosphere and Surface," Ch. 2 in IPCC, *Climate Change 2013: The Physical Science Basis*, Working Group I Contribution to the Fifth Assessment Report of the Intergovernmental Panel on Climate Change (New York, NY: Cambridge University Press, 2013), available at: http://www.climatechange2013.org/images/report/WG1AR5_Chapter02_FINAL.pdf

[6] "Detection and Attribution of Climate Change: from Global to Regional," Ch. 10 in IPCC, *Climate Change 2013: The Physical Science Basis*, Working Group I Contribution to the Fifth Assessment Report of the Intergovernmental Panel on Climate Change (New York, NY: Cambridge University Press, 2013), available at: http://www.climatechange2013.org/images/report/WG1AR5_Chapter10_FINAL.pdf

Such nuanced, carefully expressed conclusions in shades of grey do not lend themselves to effective translation when politics deals in black and white.

Are Extreme Precipitation and Flooding the Same Thing?

A common confusion is that an increase in "extreme precipitation" necessarily implies or is directly associated with an increase in flooding. This is incorrect.

Setting aside uncertainties in detection and attribution and postulating that extreme precipitation has increased and can be attributed in some part to greenhouse gas emissions does not lead automatically to findings of corresponding increases in streamflow (floods) or damage.

How can this be?

Think of it like this: Precipitation is to flood damage as wind is to windstorm damage. It is not enough to say that it has become windier to make a connection to increased windstorm damage—you would need to show a specific increase in those specific wind events that actually cause damage. There are a lot of days that could be windier with no increase in damage; the same goes for precipitation.

Even though there have been increases in what scientists call "extreme precipitation" there is very little evidence to suggest that these increases have been accompanied by increasing floods. This is a robust finding across the literature, and further details are provided below in the discussion of floods.

Absent an increase in peak streamflows caused by increasing extreme precipitation, it is impossible to find a causal linkage between increasing precipitation and increasing floods, much less between precipitation and flood damage. There are of course good reasons why a linkage between increasing precipitation and peak streamflow would be difficult to make, such as the seasonality of the increase in rain or snow, the large variability of flooding, and the human influence on river systems. Those difficulties of course translate directly to a difficulty in connecting the effects of increasing greenhouse gases in the atmosphere to flood disasters.

Let's now turn to the most damaging events, tropical cyclones, which are also among the most studied type of extreme event.

Tropical Cyclones

Hurricanes are "tropical cyclones" that occur in the North Atlantic and in the Eastern Pacific, often off the coast of Mexico. Tropical cyclones have different names in other parts of the world, like typhoon or cyclone, but they are all the same phenomena. Tropical cyclones are responsible for some of the greatest economic and human impacts from any type of extreme event. In 2005 Hurricane Katrina led to damage of more than $80 billion and in 1991 a tropical cyclone led to more than 138,000 deaths in Bangladesh.[7]

In a recent paper we found that U.S. hurricanes are responsible for almost 70% of the overall increase in disaster losses since 1980 in the Munich Re global loss dataset.[8] Consequently, if we can understand what is

---

[7] "NOAA's Top Global Weather, Water and Climate Events of the 20th Century," NOAA Backgrounder, available at: http://www.noaanews.noaa.gov/stories/images/global.pdf

[8] S. Mohleji and R. Pielke, Jr., "Reconciliation of Trends in Global and Regional Economic Losses from Weather Events:

behind that increase we will have explained a majority of recent increases in the costs of disasters.

In 2008, a team of six researchers (including me) published a paper that asked an apparently simple question: If each hurricane season of the past took place with the level of development on the nation's coasts of 2005, how much damage would have occurred in each year?

Figure 5.1: Normalized U.S. Hurricane Damage (1900-2013)

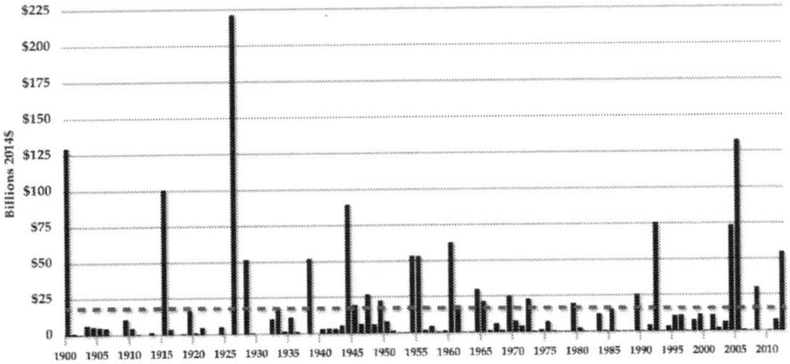

Source: Updated from R. Pielke, Jr., J. Gratz, C. Landsea, D. Collins, M. Saunders, and R. Musulin, "Normalized Hurricane Damage in the United States: 1900-2005," *Natural Hazards Review* 9 (2008): pp. 29-42.

To answer this question we took a time series of loss data kept by the U.S. National Hurricane Center from 1900 and adjusted it for inflation, a measure of household wealth, property in coastal counties and population.[9] We used two different methods which arrived at substantially similar results. The graph above shows the

---

1980–2008," *Natural Hazards Review* 15 (2014), available at: http://dx.doi.org/10.1061/(ASCE)NH.1527-6996.0000141

[9] R. Pielke, Jr., J. Gratz, C. Landsea, D. Collins, M. Saunders, and R. Musulin, "Normalized Hurricane Damage in the United States: 1900–2005," *Natural Hazards Review* 9 (2008): pp. 29-42.

results of this study, updated through 2013.[10] We call the adjusted data "normalized hurricane damage" to reflect the fact that it has been adjusted to a common base year.

The graph includes Superstorm Sandy, which by some measures was not technically a hurricane but a "post-tropical cyclone of hurricane strength." There were also three other storms since 1900 which made landfall as "post-tropical cyclones of hurricane strength" which occurred in 1904, 1924 and 1925. We don't have loss data for these three storms, so they appear in the dataset as placeholders at $5 billion each.[11]

Sandy made 2012 a bad year, but since 1900 we estimate that 8 other years would have had greater damage. The most costly year, in terms of normalized damage, was 1926 with more than $200 billion in estimated damage. This was mainly from the Great Miami Hurricane, which would devastate Miami today.

But how do we know if our estimates are any good? Maybe our methods are flawed or we have neglected to include important variables, like changing building practices, as described in Chapter 2.

We perform several independent checks on the analysis. One is simple. We know which years had exceptionally large losses: 1900 and 1915 in Galveston, 1926 and 1928 in Florida, 1938 in New England, Hugo in 1989, Andrew in 1992, and so on. These years show up clearly in our dataset with big losses.

A far more sophisticated check is to compare trends in the incidence of hurricanes with trends in damage. Because counts of hurricanes and measures of their

---

[10] You can explore this data and use it to place current hurricanes into historical context: http://icatdamageestimator.com

[11] Their inclusion does not alter this analysis.

strength are independent of damage estimates, they can serve as a basis for evaluating the appropriateness of our adjustments. Logically, we would expect that trends in normalized damage and trends in hurricane incidence would go in the same direction. It turns out that they do match up, almost perfectly.

Figure 5.2: U.S. Hurricane Landfalls (1900-2013)

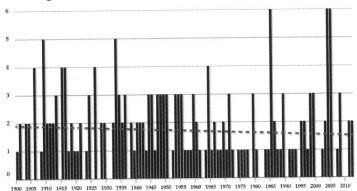

Source: NOAA/NHC.

The graph above shows a count of U.S. hurricane land-falls from 1900.[12] It shows no evidence indicating that hurricane landfalls have increased since 1900, a finding that holds if one starts the analysis in 1851 (when NOAA's dataset begins) or 1950. There is an upwards trend if the count is arbitrarily started in 1970, the lowest period of activity since 1900.

---

[12] "Atlantic Basin Comparison of Original and Revised HURDAT," Hurricane Research Division, National Oceanic and Atmospheric Administration website (Feb. 2014), available at: http://www.aoml.noaa.gov/hrd/hurdat/comparison _table.html

But that is storm frequency. What about storm intensity, in particular, the strength of storms when they make landfall?

The graph below shows the U.S. landfall intensity data for 1900 through 2013. There is no upwards trend since 1900, consistent with the trends in normalized losses. There is similarly no upwards trend in the data since 1950, but there is if the analysis is started in 1970, as is the case with landfall frequency.

Figure 5.3: Intensity of U.S. Hurricane Landfalls by Year (1900-2013)

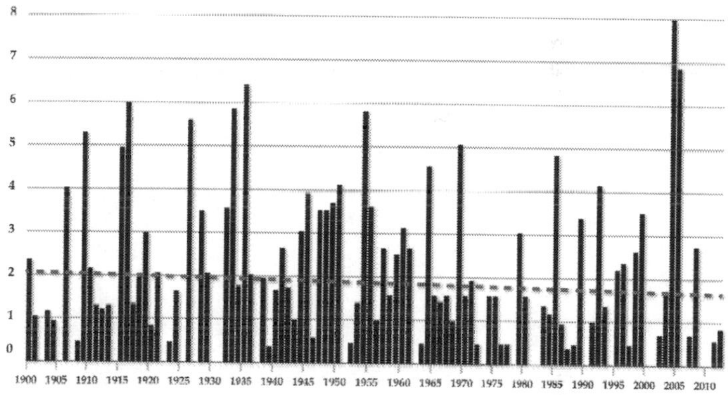

Source: NOAA, provided courtesy of C. Landsea, NHC. The vertical axis is expressed as an index.

The data show that hurricanes have not increased in the U.S. in frequency, intensity, or normalized damage since at least 1900. The trends across these three datasets match up well. Based on this match here is what we concluded in our 2008 paper:

The lack of trend in twentieth century normalized hurricane losses is consistent with what one would expect to find given the lack of trends in hurricane frequency or intensity at landfall. This finding should add some confidence that, at least to a first degree, the normalization approach has successfully adjusted for changing societal conditions. Given the lack of trends in hurricanes themselves, any trend observed in the normalized losses would necessarily reflect some bias in the adjustment process, such as failing to recognize changes in adaptive capacity or misspecifying wealth. That we do not have a resulting bias suggests that any factors not included in the normalization methods do not have a resulting net large significance. [13]

But what about the rest of the world?

The IPCC SREX report concluded "There is low confidence in any observed long-term (i.e., 40 years or more) increases in tropical cyclone activity (i.e., intensity, frequency, duration), after accounting for past changes in observing capabilities." [14] The IPCC AR5 reaffirmed this conclusion:

Current datasets indicate no significant observed trends in global tropical cyclone frequency over the past century…. In summary, this assessment does not revise the SREX conclusion of low confidence that any reported long-term (centennial) increases in tropical cyclone activity are robust, after accounting for past changes in observing capabilities. [15]

---

[13] The lack of bias includes the potential effects of sea level rise. Any effect of seal level rise on damages since 1900 is not detectable.

[14] IPCC, *Managing the Risks of Extreme Events and Disasters to Advance Climate Change Adaptation* (New York, NY: Cambridge University Press, 2012), available at: http://www.ipcc-wg2.gov/SREX/

[15] "Observations: Atmosphere and Surface," Ch. 2 in IPCC, *Climate Change 2013: The Physical Science Basis*, Working Group I Contribution to the Fifth Assessment Report of the Intergovernmental Panel on Climate Change (New York, NY: Cambridge University Press, 2013), available at:

The IPCC did find that tropical cyclone activity has increased in the North Atlantic since 1970. As shown above, U.S. normalized losses, landfall frequency and intensity have also increased in the U.S. over this period. However, the IPCC concluded that these increasing losses are a matter of the choice of the start date for the analysis, as the trends since 1970 do not exceed documented variability:

> *No robust trends in annual numbers of tropical storms, hurricanes and major hurricanes counts have been identified over the past 100 years in the North Atlantic basin.* [16]

The IPCC SREX agrees with respect to observed damage:

> *Most studies related increases found in normalized hurricane losses in the United States since the 1970s (Miller et al., 2008; Schmidt et al., 2009; Nordhaus, 2010) to the natural variability observed since that time (Miller et al., 2008; Pielke Jr. et al., 2008). Bouwer and Botzen (2011) demonstrated that other normalized records of total economic and insured losses for the same series of hurricanes exhibit no significant trends in losses since 1900.* [17]

In 2012, I was part of a research project that looked at trend in the number of tropical cyclones which made landfall around the world. [18] We found that there have

---

http://www.climatechange2013.org/images/report/WG1AR5_Chapter02_FINAL.pdf

[16] Ibid.

[17] IPCC, *Managing the Risks of Extreme Events and Disasters to Advance Climate Change Adaptation* (New York, NY: Cambridge University Press, 2012), available at: http://www.ipcc-wg2.gov/SREX/

[18] J. Weinkle, R. Maue, and R. Pielke, Jr., "Historical Global Tropical Cyclone Landfalls,*" *Journal of Climate* 25 (2012): pp. 4729–4735, available at: http://dx.doi.org/10.1175/JCLI-D-11-00719.1

been no significant trends (up or down) in global tropical cyclone landfalls since 1970 (when data allow for a comprehensive perspective; we have data going back further for various parts of the world), or in the overall number of tropical cyclones. The graph below shows those data, updated through 2013.

Figure 5.4: Global Tropical Cyclone Landfalls (1970-2013)

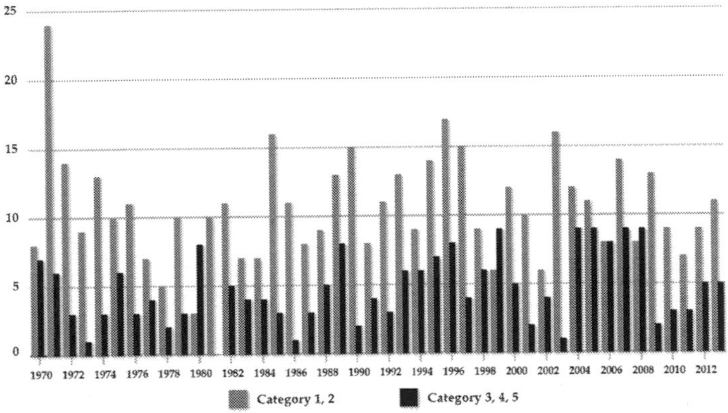

Source: After J. Weinkle, R. Maue, and R. Pielke, Jr., "Historical Global Tropical Cyclone Landfalls, *Journal of Climate* 25, No. 13 (2012): pp. 4729-4735. Thanks to R. Maue for the updated data.

Our collaborator Ryan Maue has analyzed data on total tropical cyclone activity worldwide since 1970 (not just the ones which make landfall).[19] Those data can be seen in the graph below.

---

[19]After R.N. Maue, "Recent historically low global tropical cyclone activity," *Geophysical Research Letters* 38 (2011): L14803, doi:10.1029/2011GL047711.

Figure 5.5: Total Count of Tropical Cyclones of Tropical Storm (Top Curve) and Hurricane Strength, 12-Month Running Sums (1970 - 31 Mar. 2014)

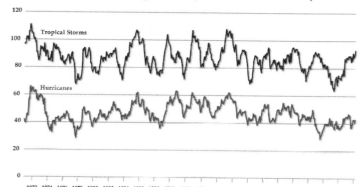

Source: R. Maue.

There is simply little evidence to support claims that tropical cyclones, or hurricanes, have become more common or intense on climate timescales, a conclusion that is strongest for landfalling storms. It is thus no surprise that normalized loss studies have also failed to find increasing trends.[20]

A Final Note on U.S. Hurricanes

As of this writing, over halfway through the 2014 hurricane season, the United States is currently in a remarkable stretch with no major hurricane (Category 3+)

---

[20] Sea level rise is ongoing and expected to continue, with potentially large impacts, according to the IPCC. However, to date there are no studies which have identified a signal of sea level rise in storm damage statistics. This is no doubt due to the fact that sea level rise to date has been relatively modest, coastal infrastructure is built over time, and coastlines are heavily managed via engineering works such as reclamation and beach nourishment.

landfalls, as shown in the figure below. The last major hurricane to strike the U.S. was Wilma back in 2005. That streak may have ended before you read this, or it might be still ongoing. Either way, it has been a remarkable period with no major U.S. hurricane landfalls.

The five-year period ending 2013 has seen 2 total hurricane (Cat 1+) landfalls. That was a record low for any five-year period since 1900. Two other five-year periods have seen 3 landfalls (years ending in 1984 and 1994). Prior to 1970 the fewest landfalls over a five-year period was 6. From 1940 to 1957, every 5-year period had more than 10 hurricane landfalls (1904-1920 was almost as active). These data suggest that the U.S., even with Superstorm Sandy, has been in a relatively benign period of hurricane activity, at least as compared to past eras.

Figure 5.6: Days Between Major Hurricane Landfalls in the U.S. Since 1900

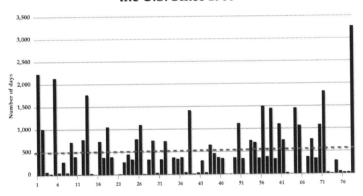

There were 78 major hurricane landfalls in the U.S. from 1900 to 2013. Source: NOAA.

The Bottom Line on Tropical Cyclones

With respect to the central question that is the focus of this short volume, the data are clear. Those who assert that disasters are getting more costly because of climate change (human caused or not) are going to have to look at phenomena other than U.S. hurricanes or tropical cyclones around the world. There is no evidence to suggest that hurricanes have become more common, intense or costly for any reason other than more people and their property exist in locations vulnerable to the impacts of these powerful storms.

Floods

As with tropical cyclones, there is little evidence in support of claims that floods have become more common or more intense. The IPCC AR5 concludes:

In summary, there continues to be a lack of evidence and thus low confidence regarding the sign of trend in the magnitude and/or frequency of floods on a global scale. [21]

The IPCC SREX went into a bit more detail:

- There is limited to medium evidence available to assess climate-driven observed changes in the magnitude and frequency of floods at regional scales.

---

[21] "Observations: Atmosphere and Surface," Ch. 2 in IPCC, *Climate Change 2013: The Physical Science Basis*, Working Group I Contribution to the Fifth Assessment Report of the Intergovernmental Panel on Climate Change (New York, NY: Cambridge University Press, 2013), available at: http://www.climatechange2013.org/images/report/WG1AR5_Chapter02_FINAL.pdf

- There is low agreement in this evidence, and thus overall low confidence at the global scale regarding even the sign of these changes.[22]

In early 2014, more than a dozen contributors to the IPCC SREX published a peer-reviewed paper specifically on floods, to expand their discussion of the state of the science.[23] The authors are clear that they are speaking for themselves, and not the IPCC, but they explain that their conclusions are "congruent" with those found in the SREX. Here are a few of their conclusions, which they note were focused on riverine floods, and not storm surges caused by coastal storms.

The authors explain: "a direct statistical link between anthropogenic climate change and trends in the magnitude/frequency of floods has not been established." More specifically,

> [N]o gauge-based evidence has been identified for a clear climate-driven, globally widespread, observed change in the magnitude/frequency of river floods during the last decades. There is thus low confidence regarding the magnitude/frequency and even the sign of these changes.

One important reason for this conclusion is that there are limited time-series data available in many regions of the world.

---

[22] IPCC, *Managing the Risks of Extreme Events and Disasters to Advance Climate Change Adaptation* (New York, NY: Cambridge University Press, 2012), available at: http://www.ipcc-wg2.gov/SREX/

[23] Z.W. Kundzewicz, S. Kanae, S.I. Seneviratne, J. Handmer, N. Nicholls, P. Peduzzi, R. Mechler, L.M. Bouwer, N. Arnell, K. Mach, R. Muir-Wood, G.R. Brakenridge, W. Kron, G. Benito, Y. Honda, K. Takahashi, and B. Sherstyukov, "Flood risk and climate change: global and regional perspectives," *Hydrological Sciences Journal* 59 (2014): pp. 1-28.

The authors are explicit in noting that trends in "extreme precipitation" have not translated to increased riverine flooding:

> *Despite the diagnosed extreme-precipitation-based signal, and its possible link to changes in flood patterns, no gauge-based evidence had been found for a climate-driven, globally widespread change in the magnitude/frequency of floods during the last decades.*

The authors conclude their analysis with a plea to focus attention on more important issues than establishing a linkage between greenhouse gases and flood trends:

> *There is such a furor of concern about the linkage between greenhouse forcing and floods that it causes society to lose focus on the things we already know for certain about floods and how to mitigate and adapt to them. Blaming climate change for flood losses makes flood losses a global issue that appears to be out of the control of regional or national institutions. The scientific community needs to emphasize that the problem of flood losses is mostly about what we do on or to the landscape and that will be the case for decades to come.*

Amen.

Tornadoes

Tornadoes occur in many regions around the world, but are most commonly found in North America. I recently was part of a research team that sought to apply the normalization methods that we first developed for U.S. hurricanes to tornadoes.[24]

---

[24] K.M. Simmons, D. Sutter, and R. Pielke, Jr., "Normalized tornado damage in the United States: 1950-2011," *Environmental Hazards* 12 (2013): pp. 132-147.

Figure 5.7: Normalized U.S. Tornado Damage
(1950-2013)

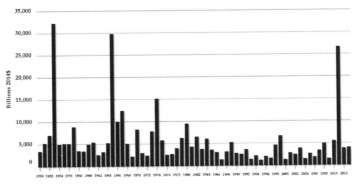

Updated courtesy K. Simmons.

The figure above shows an estimate of how much tornado damage would occur in the United States, if each year's tornadoes occurred with today's levels of population and development. The worst year for damage was 1953, but 1965 and 2011 aren't far behind. In terms of loss of life, 2011, with 560 deaths, saw the most casualties since 1925, when 794 people died. Overall, however, the United States has seen a long-term decrease in both property damage and loss of life related to tornadoes. Yet even with this decline, 2011 reminds us that large impacts are always possible.

The graph reflects data from almost 58,000 tornadoes from 1950 through 2013. Using damage estimates of the U.S. government's National Oceanic and Atmospheric Administration, we used several approaches to normalize the losses to 2014 values in order to estimate how much damage would occur if historical tornadoes occurred with today's levels of population and development. The figure above shows our central estimate.

As with our work on hurricanes, a first question to ask is: how do we know if our estimates are any good?

And as with hurricanes we perform several independent checks. One is that we know which years had exceptionally large losses: 1953, 1965, 1974 and 2011. These four years show up clearly in our dataset as outliers.

A more sophisticated check is to compare trends in the incidence of tornadoes with trends in damage. Because counts of tornadoes are independent of damage estimates, they can serve as a basis for evaluating the appropriateness of our adjustments. Logically, we would expect that trends in damage and trends in tornado incidence would go in the same direction. This check is a bit tricky because meteorologists have changed how they track tornadoes over time, requiring us to break the overall dataset into a series of shorter periods. When we do this, we find excellent agreement between the damage trends and the trends in tornado incidence, giving us some confidence in our approach.

Average annual losses for the entire 63-year period across the U.S. are $5.9 billion (in 2014 dollars). However, for the first 32 years of the dataset (1950-1981) the annual average was $7.6 billion and since 1982, a period of 31 years, the annual average has been $4.1 billion, a drop of more than 50%.

Does the drop in average annual damage mean that there have actually been fewer tornadoes? Not necessarily.

The IPCC SREX explains that the quality of the data makes any conclusions about long-term trends problematic: "There is low confidence in observed trends in small spatial-scale phenomena such as tornadoes and hail."[25] In our analysis we concluded that the data are

[25] IPCC, *Managing the Risks of Extreme Events and Disasters to Advance Climate Change Adaptation* (New York, NY: Cambridge

"suggestive" of an actual decline in tornado incidence, but do not say anything stronger, and recommend further research.

Drought

As with tropical cyclones, floods, and tornadoes, there is little evidence to support claims that drought has increased globally on climate time scales.

The IPCC SREX concluded:

There is medium confidence that since the 1950s some regions of the world have experienced a trend to more intense and longer droughts, in particular in southern Europe and West Africa, but in some regions droughts have become less frequent, less intense, or shorter, for example, in central North America and northwestern Australia.[26]

For the U.S., the National Climate Assessment (NCA) concluded: "There has been no universal trend in the overall extent of drought across the continental U.S. since 1900."[27] The NCA did point to regional differences, with some regions experiencing more periods of drought, such as the U.S. Southwest, and others less, like the Midwest.

The IPCC recently summarized its findings on drought, reaching a conclusion of "low confidence" on either detection or attribution:

University Press, 2012), available at: http://www.ipcc-wg2.gov/SREX/

[26] Ibid.

[27] See the National Climate Assessment report at: http://nca2014.globalchange.gov/; similarly, the U.S. NCA did not find evidence of an nationwide increase in floods, tornadoes or landfalling hurricanes, perfectly consistent with the IPCC and the information presented here.

> *There is not enough evidence to support medium or high confidence of attribution of increasing trends to anthropogenic forcings as a result of observational uncertainties and variable results from region to region (Section 2.6.2.3). Combined with difficulties described above in distinguishing decadal scale variability in drought from long-term climate change we conclude consistent with SREX that there is low confidence in detection and attribution of changes in drought over global land areas since the mid-20th century.[28]*

In addition, the IPCC also concluded that recent drought in the western United States—that highlighted by the U.S. NCA—could not be attributed to human-caused climate change:

> *Recent long-term droughts in western North America cannot definitively be shown to lie outside the very large envelope of natural precipitation variability in this region (Cayan et al., 2010; Seager et al., 2010), particularly given new evidence of the history of high-magnitude natural drought and pluvial episodes suggested by paleoclimatic reconstructions.*

There is little evidence provided by the IPCC to support claims that drought has become more frequent globally on climate timescales. Further, there is also little evidence in support of claims of attribution of causes for trends in regional drought.

At the regional level consider, for example, property damages resulting from bushfire in Australia. Bushfire

---

[28] "Detection and Attribution of Climate Change: from Global to Regional," Ch. 10 in IPCC, *Climate Change 2013: The Physical Science Basis*, Working Group I Contribution to the Fifth Assessment Report of the Intergovernmental Panel on Climate Change (New York, NY: Cambridge University Press, 2013), available at: http://www.climatechange2013.org/images/report/WG1AR5_Chapter10_FINAL.pdf

disasters are, unfortunately, a common occurrence in Australia and like many extremes they are often attributed to human-caused climate change.

In 2009 I was part of a team of researchers who applied a normalization technique to historical losses from bushfires.[29] You can see the results of that research in the next set of graphs.

The top panel shows how many buildings have been destroyed in bushfires from 1926 to 2009. However, over that time period there have been many more buildings built in locations prone to bushfires. The bottom panel shows losses after they have been normalized to 2009. There is no evidence in the normalized data of an increasing trend in losses. In the logic of Figure 3.1, neither detection nor attribution has been achieved.[30]

Normalization based simply on changes in number of buildings of course does not provide a complete picture of societal changes during this period (for example, we did not try to estimate changes in the fire-resistance of buildings). Nonetheless, the normalized data *are* consistent with known patterns of climate variability, specifically the El Niño Southern Oscillation and the Indian Ocean Dipole. That is, after normalization, we see a good correspondence between bush fire losses and climate variability. As I discussed earlier, when normalized social trends match up well with independently

---

[29] R.P. Crompton, K.J. McAneny, K.P. Chen, R.A. Pielke, and K. Haynes, "Influence of Location, Population, and Climate on Building Damage and Fatalities due to Australian Bushfire: 1925-2009," *Weather, Climate, and Society* 2 (2009): pp. 300-310.

[30] That paper prompted a response from Neville Nicholls and a subsequent rejoinder from us. I encourage anyone interested in this subject to read all three papers, which can be found at: http://sciencepolicy.colorado.edu/publications/special/austr alian_bushfires.html

observed climate trends, it gives us confidence that our normalization assumptions are reasonable.

Figure 5.8: Building Damage from Australian Bushfires (1925-2008)

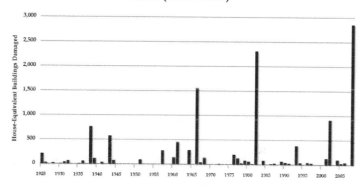

Normalized Building Damage from Australian Bushfires (1925-2008)

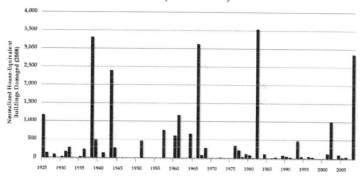

Source: R.P. Crompton, K.J. McAneny, K.P. Chen, R.A. Pielke, and K. Haynes, "Influence of Location, Population, and Climate on Building Damage and Fatalities due to Australian Bushfire: 1925-2009," *Weather, Climate, and Society* 2 (2009): pp. 300-310.

Conclusion

This short volume has sought to answer a straight-forward question:

Have disasters become more costly because of human-caused climate change?

Only one answer to this question is strongly support-ed by the available data, the broad scientific literature, and the assessments of the IPCC:

No. There is exceedingly little evidence to support claims that disasters have become more costly because of human-caused climate change.

Of course, a lack of evidence does not prevent people from believing in God, aliens, or for that matter, a small celestial teapot orbiting the Sun in the asteroid belt. People may indeed have very good reasons for believing in any of these things for which data and observational evidence are unsupportive, unavailable, or inconclusive. The issue of disasters and climate change will be no different.

And of course, science evolves. There may be future research which overturns present understandings. If and when that happens, our assessment of what the science says should change accordingly. Thus, the conclusions presented here should be interpreted as an indication of the current state of scientific understandings, and not a prediction of what a future scientific assessment might say in the years to come.

Nonetheless, one point should be abundantly clear. The evidence available today points to a clear answer to the central question at the focus of this short volume: Human-caused climate change has *not* led to a detecta-ble increase in the costs of disasters.

But the climate is changing. It would be a mistake to conclude that because the evidence shows that human-caused climate change has not led to demonstrable increases in the costs of disasters that (a) climate change is not occurring, or (b) we need not worry about it.

In this regard, those advocates for action who claim to see the influence of climate change do themselves no favors by stretching and sometimes going beyond what science can support. Sure, you can get attention and news coverage with assertions that changes in climate are leading to more disasters. But over the long term, are such strategies worth the risk of exaggerating what science can actually show with evidence?

That the Administration can state scientific conclusions which are clearly not supported, even by its own scientific assessment, without fear of challenge by the scientific community or the media suggests that the climate science community appears to have succumbed to "noble cause corruption." The ends seem to justify the means.[4]

But whether you believe the spin or the science, opinion polls have not changed as science advisor Holdren predicted. In fact, as I document in *The Climate Fix*, public opinion on climate has been remarkably stable over decades, even as the absolute cost of disasters has increased.

Apocalyptic visions are a bit like addictive drugs. Upon repeated usage, the dosage needs to be upped to achieve the same effect. In this way, efforts to politicize connections between greenhouse gases and extreme events have a tendency to go well beyond what science can support. With fervent advocates ready to attack anyone who steps out of line, as they did when I wrote for *FiveThirtyEight*, there can be significant obstacles for independent experts to weigh in when claims are made well beyond that which science can support.

The goal of political debate of course should not be to get everyone to think alike, but rather, to secure collective action in common interests. Often lost in the passions of

---

[4] With the challenges to the president's claims coming almost exclusively from those opposed to his political agenda on climate, it reinforces the politicization of the science of climate change and extreme events. When science advisor Holdren was challenged under the so-called Data Quality Act by a conservative group on one of his public claims about the association of extreme winter weather of 2014 and climate change, the White House responded by explaining that Holdren's comments represented personal opinion rather than scientific conclusions. The exchange went unreported by the major media.

the climate debate is that the world is not making progress towards reducing the emission of greenhouse gases.

More to the point, if reducing GHG emissions is going to be an important part of our strategy for confronting climate change, then one can only acknowledge that, after more than 25 years of efforts to craft national and international climate policies, this strategy has been an abject failure.

While battles over the news cycle rage on, ultimately, a more effective approach to climate policy will inevitably focus on listening to the public, rather than trying to trick them by exaggerating or distorting the work of authoritative experts.

This volume concludes with a broader look at climate policy, well beyond the issue of disasters and climate change, with a positive focus on those steps more likely to bear fruit in the challenge of stabilizing concentrations of carbon dioxide in the atmosphere and adapting to a changing climate.[5] Thus, this concluding chapter goes well beyond the issue of disasters and climate change, to engage a bit more deeply with the "so what?" question.

The Only Equation You Need to Know

To understand efforts to stabilize the amount of carbon dioxide in the atmosphere requires a basic understanding of where carbon dioxide comes from and the policy tools which can be used to reduce emissions. A very simple but powerful framework for such an understanding was proposed in the 1980s by a Japanese scientist, Yoichi Kaya, as

---

[5] The following section draws on and updates an essay that first appeared in *Foreign Policy* in 2012. That piece was in turn based on the much more comprehensive analysis found in *The Climate Fix* (Basic Books, 2011).

a tool for creating scenarios of future emissions that would be used as inputs to climate models. If you want to model the future climate, you would of course need to know something about future rates of emissions, which are primarily the result of the burning of fossil fuels — coal, natural gas, and petroleum.

Kaya explained that future carbon dioxide emissions would be the product of four factors: population, economic activity, how we obtain our energy, and how we use that energy. From a policy perspective, these four factors can be used to describe in totality the means available to reduce future carbon dioxide emissions. In short, our levers available to reduce emissions are population, income, and energy consumption or production. Those are the tools in the tool box.

We can simplify these four factors even further — population and income together are simply GDP, or aggregate economic activity, and the production and consumption of energy is a reflection of the energy technology that we use (measured as carbon dioxide released per unit of GDP).

So the Kaya Identity — as it has come to be called — simply says that:

*Emissions = GDP * Technology*

With this simple identity before us, right away we should see a fundamental challenge to reducing emissions, because a rising GDP, all else equal, means more emissions.

If there is one ideological commitment that unites nations and people around the world in the early 21st century, it is that GDP growth is non-negotiable. Currently policy makers on six different continents are focused on efforts to grow GDP, and with it jobs and wealth.

The Iron Law

If you spend any time in the midst of the climate debate, it won't be long before you will be assailed by those who would like to argue that economic growth is unnecessary or even undesirable. I hear these arguments mostly from economically comfortable academics in posh university towns across the richer parts of the world. Without dismissing the pedagogical importance of such debates, it is quite easy to observe that no political candidate (much less a government) has ever secured political office on a platform of de-growth. One has to truly live in an insulated bubble to think that any such policies will be used as a response to climate change.

I have called such dynamics the "iron law of climate policy." The law says that while evidence shows that people around the world are willing to pay some price to attain environmental objectives (and that willingness differs in different places, of course), such willingness is limited. When you consider that stabilizing the amount of carbon dioxide in the atmosphere at a low level (like the oft-quoted 450 parts per million) requires cuts in emissions of 50% (or much more) in coming decades, it should be clear that reducing GDP is never going to be an effective tool of climate policy.

So the Kaya Identity tells us that the focus of policy to stabilize carbon dioxide in the atmosphere necessarily must be on technology, not on reducing GDP, and here the math is surprisingly simple. To achieve a stabilization of carbon dioxide in the atmosphere (at any level) requires that more than 90% of the energy that we consume comes from carbon-free sources, like nuclear, wind, or solar, or even coal or gas with carbon capture and storage. This 90% threshold is independent of how much energy the world consumes, which in round numbers is in the future going to be a lot more than today, especially when you consider that two billion people or more lack even basic access to energy. The time that it takes to reach that 90% threshold will determine the level at which carbon dioxide is stabilized. The sooner we do so, the lower the level.

The figure below uses data from the energy company BP and shows that in 2012, the world obtained about 13% of its energy consumption from carbon-free sources. That 13% proportion has not increased in about two decades. It had doubled from 1965 to 1990, primarily due to the dramatic expansion of nuclear power. If stabilization of carbon dioxide is to be achieved, that 13% proportion must increase to above 90%, regardless how much energy we eventually consume. That is the hard math underlying the climate debate, a math that is obscured by the public wars over who gets to be a legitimate voice in the climate debate.

Figure 6.1: Carbon-Free Energy in the Global Mix (1965-2012)

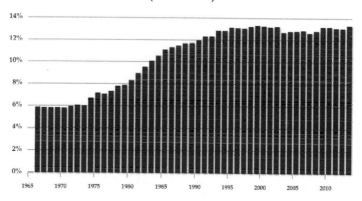

Source: BP.

Is going from 13% to 90% a big deal? What is the size of the technological challenge in more intuitive terms?

It is a big deal. In a nice round number, if the goal is to achieve stabilization by 2050, then the world would need to deploy the equivalent of a nuclear power plant-worth of carbon-free energy every day between now and 2050, while retiring an equivalent amount of fossil generation. If you prefer to use wind or solar as a measuring stick the numbers are just as mind boggling—about 1,000 wind turbines per day, or 250 solar thermal plants.[6].

That one-nuke-plant-per-day assumes very low levels of growth in energy access around the world. More rapid expansion of modern energy consumption worldwide implies a need for greater rate of deployment of carbon-free energy.

What Specialists Argue About

Many specialists in the climate debate understand both the iron law and the magnitude of the technological challenge. Thus, the informed debate over climate policy revolves around competing strategies to motivate technological innovation in the production and consumption of energy. For many years the dominant view was that a price on carbon—whether through a tax or a traded permit system—would be necessary and perhaps even sufficient to create economic incentives to stimulate innovation.

Whatever economic theory may say about the merits of carbon pricing, the model has repeatedly failed the most basic of real-world political tests.[7] That this failure is inev-

---

[6] These (round) numbers come from the calculations found in Chapter 4 of *The Climate Fix*, and refer to a 2.5 megawatt turbine and a 10 megawatt solar thermal plant, each operating at 30% efficiency. As I explain there, these numbers are measuring sticks: they ignore technological deployment issues like storage and transmission.

[7] J. D. Jenkins, "Political economy constraints on carbon pricing policies: What are the implications for economic efficiency, envi-

itable derives not from economics but politics. The logic of carbon pricing is that higher-priced energy would create an incentive to use it more sparingly and to invest in innovation necessary to generate cheaper alternatives. The incentive would be strengthened over time in a manner consistent with some timetable for hitting an emissions reduction target. The seductive logic of carbon pricing has gained it many adherents, especially among academics and scientists, and among many rich-world environmentalists.

From a political perspective, in the real world what economists call "incentives" would be more properly called "economic pain" and a "strengthening of incentives" would be called "turning the screws." Any policy which depends for its success on creating economic stress on consumers (who are also voters in many places) to motivate change, is a policy doomed to fail. Voters will respond of course, typically by voting out of office any politician or party who is perceived to be working against their economic interests. At the same press conference that announced President Obama's new commitment to climate change, his spokesman explained: "We have never proposed or supported a gas tax."[8] Of course they haven't: it would be political suicide.

Supporters of carbon pricing have no good answer for the politics. Claims that a carbon price won't even be noticed in the economy belie the fundamental logic of the approach, which is to create a powerful incentive for

---

ronmental efficacy, and climate policy design?" *Energy Policy* 69 (2014): pp. 467-477.

[8] "Daily Briefing by Press Secretary Jay Carney and Senior Advisor to the President John Podesta 05/05/14," The White House, Office of the Press Secretary (5 May 2014), available at: http://www.whitehouse.gov/the-press-office/2014/05/05/daily-briefing-press-secretary-jay-carney-and-senior-advisor-president-j

change. Calls for a price high enough to actually motivate profound change are so unrealistic as to be laughable. In Australia, the introduction of a carbon tax was accompanied by tax reform that subsidized the tax—that is, the government is redistributing to consumers more money than collected by the tax. Still the policy remained unpopular, with more than 60% opposed to the tax, despite similar numbers supporting action on climate. The unpopular tax arguably played a role in the fall of Julia Gillard, now a former prime minister, the subsequent electoral loss by Labor, and ultimate repeal of the tax.

If debate over the economics and politics of carbon pricing took place only at the theoretical level, the discussion might never end. But evidence from the real world provides a broad basis for evaluating the prospects for carbon pricing and its effects on both emissions and innovation. That evidence shows that in the few places where a carbon price has been implemented, there is no evidence of changes in the technology of energy production and consumption that are even remotely close to the magnitude required to stabilize emissions.

Europe has provided the richest body of experience for evaluating carbon pricing, and the evidence shows that not only has the strategy done very little compared with what would be necessary, but that major economies within Europe have in recent years become more, not less, carbon intensive. Germany in particular has experienced a "dash for coal." The German government, having expressed a desire to shut down nuclear power and fossil fuel power, is quickly coming to the realization that such ambitions are not presently possible. In fact, Germany is building new coal power plants. The flawed design of the European carbon trading system has prompted many calls for wholesale reform.

Why the Climate Wars are so Angry

Public debate on carbon pricing has typically segregated itself according to different perspectives on the costs and benefits of action. Some argue that the century-long costs of inaction are greater than the costs of action, others take the opposite stance. The fact is that many aspects of this debate simply cannot be resolved through evidence, as we don't have any actual data about the future, just assumptions. This means that combatants often replace reasoned policy debates with appeals to authority, ad hominem attacks, and other tactics that have the effect of making the climate debate particularly nasty and destructive.

With respect to costs, advocates for carbon pricing typically argue that the costs of action are low, or even somewhat implausibly, cost-free. The typical tactic is to use an economic model to project net costs over the better part of a century. Such models, laden with phantasmagoric assumptions such as the pace of future technological innovation, offer little solace to the politician who runs for election every few years. Opponents argue that higher priced energy will kill jobs and force industries to relocate. With only a few exceptions, such as hydro-power-rich British Columbia, governments around the world have decisively sided with the opponents of carbon pricing, and any carbon price put into place, such as in my home town of Boulder, is too low to have any meaningful effect on energy use. Even Australia's politicians who advocate carbon pricing have accepted that the tax would not have achieved the nation's modest emissions-reduction goals. Those goals, if they are to be achieved on paper, will have to be met through the purchase of "offsets," another curious invention designed to mute the pain caused by carbon pricing.[9]

On the benefits side of the equation, most of the debate has focused on the desirable consequences of reducing emissions. Yet such benefits will only play out in the future, decades or even a century or more from now, and the only tools available for specifying such benefits are complex computer models. Needless to say, the uncertainties surrounding the scope and magnitude of future benefits are incredibly uncertain, and the results of efforts to model them are strongly dependent on assumptions that scientists must make about both society and climate in the coming decades. Such uncertainty means that there are many plausible futures.

As in the case of the run-up to the Iraq war, which I discussed earlier, fear can be a powerful political tool. Efforts to stoke alarm have no apparent limit, with weather extremes and other bad things (such as species extinctions, spreading diseases, food crises, even civil unrest) often being linked to greenhouse gas emissions, no matter how tenuous the science.

Opponents to action do the same thing with snowstorms, cold summers, and extended periods with no hurricanes. Such tactics bear out a warning offered by the late Steve Schneider: "uncertainties so infuse the issue of climate change that it is impossible to rule out either mild or catastrophic outcomes, let alone provide confident probabilities for all the claims and counterclaims made about environmental problems."[10] Science and nature provide

---

[9] Carbon "offsets" refer to the practice of paying a price for emissions reductions made elsewhere. Often these "reductions" are in the form of reductions from a counterfactual baseline, that is, increases which occur at a lower rate than expected. Much has been written on this topic.

[10] As quoted in *The Climate Fix*.

enough varied experience to turn the future into a Rorschach test whose interpretation can reflect anyone's contemporary political viewpoint.

In such a situation, debate typically devolves into various efforts to winnow away uncertainties through methods such as identifying a consensus opinion through survey, poll, or even fiat. Yet science is stubborn in its affinity for the empirical, and uncertainties persist. If science won't yield unambiguous answers, the natural next step is to win a debate through political means. Opponents in such debates resort to proxies of expertise to try to assert some phantom mandate. The end result has been neither to win the debate nor to secure a political mandate, but to politicize the science itself.

But What About the Deniers?[11]

Conventional wisdom in the climate debate is that climate skeptics, who have more recently been promoted to climate "deniers" (thus equating them with those who deny the existence of Nazi genocide), are all-powerful forces bankrolled by rich corporations and who have wielded their awesome power to block efforts to deal with the threat of human-caused climate change. How do we know that climate skeptics have such power? As Martin Wolf of the *Financial Times* explains, it is the "world's inaction" on climate policy which shows the power of the skeptics.[12]

---

[11] This section draws upon a piece of mine first published in *The Guardian*. See R. Pielke, Jr., "Have the climate sceptics really won?" *The Guardian* (24 May 2013), available at: http://www.theguardian.com/science/political-science/2013/may/24/climate-sceptics-winning-science-policy

[12] M. Wolf, "Global inaction shows that the climate skeptics have already won," *Financial Times* (21 May 2013), available at:

From this perspective then, a key challenge of securing action on climate change is to defeat the skeptics. In the words of Lord Stern, the author of a well-known 2006 report on the economics of climate change, we must "drive back" the "forces of darkness"[13] so that the forces of good might prevail. Victory will be achieved by winning the battle for public opinion on the state of climate science.

However, a closer look at the logic underlying assertions about the power of the skeptics suggests that it is not only flawed but counterproductive. Flawed because countries where skeptics have little or no political presence, like Germany or Japan, are not doing any better at reducing emissions than countries, like the U.S., where the skeptics have been portrayed by advocates of aggressive climate action as the main reason for lack of progress. Yet the belief in the power of the skeptics leads to an ineffectual policy strategy by those who desire action; indeed, there is reason to believe that it actually works against effective action.

Advocates for action on climate change often adopt what scholars of science communication have called the "deficit model" of science. This view of the role of science in policymaking suggests that the public lacks knowledge that, if known properly (thus closing the deficit), would lead them to favor certain policy actions. The basic logic at work here is that if you only understood the "facts" as I understand them, then you would come to share my policy preferences.

---

http://www.ft.com/intl/cms/s/0/9742cc76-c142-11e2-b93b-00144feab7de.html#axzz3GLVge28C

[13] F. Harvey, "Prince Charles attacks global warming skeptics," *The Guardian* (9 May 2013), available at:
http://www.theguardian.com/environment/2013/may/09/prince-charles-climate-change-sceptics

The deficit model helps to explain why people argue so passionately about "facts" in public debates over policies with scientific components. If you believe that acceptance of certain scientific views is a precondition for or even a causal factor in determining what policy views people hold, then arguments over facts will serve as political debate by proxy. Defeat the skeptics, the logic holds, and the politics will follow.

The deficit model also helps to explain the presence of "noble cause corruption" in science, as it reinforces the idea that policy outcomes can be achieved by modulating belief.

Dan Kahan of Yale University has conducted several studies of public views on climate change and finds that the causal mechanisms of the "deficit model" actually work in reverse: data show that people typically "form risk perceptions that are congenial to their values."[14] Our political views shape how we interpret facts. Kahan's research shows that there is little if any difference in the level of scientific understanding between citizens who are very concerned about climate change and those who believe the problem is exaggerated. What differs between those groups is their political views.

On an issue as complex and data-rich as climate, there are enough data and interpretations to offer support to most any political agenda. Thus we have arguments over the degree of agreement or lack thereof among scientists and efforts to delegitimize outlier positions in order to assert one true and proper perspective. Adding to the mix is the temptation to push "facts" beyond what science can support, which offers each side opportunity for legitimate critique of the excesses of their opponents.

---

[14] D.M. Kahan, H. Jenkins-Smith, and D. Braman, "Cultural cognition of scientific consensus," *Journal of Risk Research* 14, No. 2 (2011): pp. 147-174.

In the first half of the twentieth century American political commentator Walter Lippmann recognized that uniformity of perspective was not necessary for action to take place in democracies. He explained that the goal of politics is not to make everyone think alike, but rather, to get people who think differently to act alike. A large body of scholarship supports the limitations of the deficit model, yet it remains a defining feature of debates over climate policy.

It is bad enough that those operating under the assumptions of the deficit model are wasting their time or even working against their own interests. What is worse is that such strategies fail to recognize that the battle over public opinion on climate change has long been over—it has been won, decisively in fact, by those favoring action.

Data on public opinion on climate change have been collected, in some cases for several decades, in countries around the world. What studies show is remarkably strong support for the so-called scientific consensus about the reality of climate change, as well as strong support for policy action. Even in the notoriously climate skeptical United States, the Gallup organization finds that: "trends throughout the past decade—and some stretching back to 1989—have shown generally consistent majority support for the idea that global warming is real, that human activities cause it, and that news reports on it are correct, if not underestimated."[15]

Internationally, a Gallup poll in 2007 and 2008 of 128 countries found that strong majorities of the general population in most countries—including most large emitters of carbon dioxide—believe that global warming is a

---

[15] L. Saad, "American's Concerns About Global Warming on the Rise," Gallup Politics website (8 Apr. 2013), available at: http://www.gallup.com/poll/161645/americans-concerns-global-warming-rise.aspx

result of human activities.[16] Public opinion does vary a great deal — often literally with the weather — but it has overall been remarkably consistent over many years in support of action. Public opinion, rather than an obstacle to action on climate change, is in fact a resource to be capitalized upon.

Studies of the relationship of public opinion and political action on a wide range of subjects show nothing unique or even interesting about the state of public opinion on climate change. Significant policy action has occurred on other issues with less public support on many occasions, as I have documented in *The Climate Fix*.

So public opinion should be a political asset for those working to advance policies to address climate change. But efforts to intensify public opinion through apocalyptic visions of weather-gone-wild or appeals to scientific authority, instead of motivating further support for action, have instead led to a loss of trust in campaigning scientists. Citing the ample evidence of the ineffectiveness of such approaches, Dan Kahan complains of climate campaigners, "They keep pounding the data, and with a rhetorical hammer that drives home all the symbolism that generates distrust and resistance in larger parts of the population.... Why?"[17]

---

[16] B.W. Pelham, "Awareness, Opinions About Global Warming Vary Worldwide," Gallup World website (22 Apr. 2009), available at: http://www.gallup.com/poll/117772/Awareness-Opinions-Global-Warming-Vary-Worldwide.aspx#2

[17] D. Kahan, "Annual 'new study' finds 97% of climate scientists believe in man-made climate change; public consensus sure to follow once news gets out," The Cultural Cognition Project at Yale Law School website (17 May 2013), available at: http://www.culturalcognition.net/blog/2013/5/17/annual-new-study-finds-97-of-climate-scientists-believe-in-m.html

If public opinion is not the reason we have failed to make much progress on climate change, then what is? Two of the biggest obstacles, which I have discussed at length elsewhere, are summarized here.

The first is a failure of political plausibility. As discussed above, conventional wisdom on climate policy has long been that energy prices need to increase significantly. More expensive energy fits into a complex causal chain of policy action as follows:

Win public opinion via closing the science deficit (now focused on claims about extreme weather events), defeating the skeptics →

the scientifically informed public will pressure politicians for action →

politicians respond by passing laws, and international treaties are signed →

dirty fossil energy becomes more expensive →

people consequently feel economic pain (incentives) →

not liking economic pain, people change their behavior and the market responds with more energy efficiency and fossil fuel alternatives →

such market demand stimulates innovation in the public and private sectors, as well as in civil society →

the resulting innovation delivers low carbon alternatives →

GHG emissions go down, extreme weather (and other) problems are thus solved.

Laid out from start to finish the entire causal chain seems like a Rube Goldberg invention. If the causal chain begins to weaken at the first step, where the deficit model is assumed to operate, it completely breaks apart at the point where energy is supposed to become more expensive in order to create incentives to propel efficiency and innovation. The idea that higher-priced energy can be used as a lever to transform the global energy system may work in abstract economic models, but fails spectacularly in real-world politics, where energy costs are directly linked to virtually every aspect of human well-being, from the price of food to the availability of decent jobs.

A second obstacle is the pathological obsession of many climate campaigners with the climate skeptics. By concluding that the skeptics are the main obstacle to action, campaigners are devoting their energies to a fruitless fight. Make no mistake, fighting skeptics has its benefits — it reinforces a simplistic good-versus-evil view of the world, it gives a sense of doing *something* about climate change, and elevates scientific expertise to a privileged place in policy debates.[18] However, one thing that it does not do is contribute towards effective action on climate change.

The battle over public opinion on climate change has long been won, and not by the skeptics. However, simply by virtue of their continued existence, the climate skeptics may have the last laugh, because many climate campaigners seem to be able to see nothing else in the debate. Climate skeptics are not all powerful and may not even be much relevant to efforts to decarbonize the global economy. They are not the reason that we haven't solved the climate change problem, but they are an easy explanation for more than twenty years of failed campaigning.

So What Next?

The science that shows a human impact on the planet has been convincing for many decades, at least back to the late 1980s. That impact includes but is not limited to the effects of carbon dioxide on atmospheric temperature. We are indeed running risks with the future climate through the unmitigated release of carbon dioxide into the atmos-

---

[18] I have also concluded based on my experiences that for many, participating in the climate wars serves as a fun online hobby.

phere. But unlike the debate over the costs of carbon pricing, debate over the benefits of mitigation cannot be resolved empirically, at least not on political time scales, for the simple reason that changes in climate are observed over many decades and centuries.

The great irony here of course is that the debate over the science of disasters and climate change is completely unconnected from the debate over carbon pricing—at least in the realpolitik world of economic policies—as no scenario of doom and gloom is going to convince governments to inflict short-term economic harm on the citizens who legitimize their power. Yet climate campaigners often act as if such a calculus is possible. For instance, the *New York Times* recommended exactly this strategy for Indonesia, demanding that the nation take immediate steps to reduce emissions, "sacrificing short-term economic gain for the long-term health of the planet."[19]

Although the climate issue has largely been subsumed to economic and other agendas of most of the world's policymakers, there remains a solid basis for concern about the risks of climate change due to the accumulation of carbon dioxide in the atmosphere. While the climate wars will go on—perhaps forever—characterized by a poisonous mix of dodgy science, personal attacks, and partisan warfare, the good news is that progress can yet be made outside of this battle.

The key to securing action on climate change is to break the problem down into more manageable parts. This is already happening. For instance, a coalition of activists and politicians, including numerous prominent scientists, has argued that there are practical reasons to focus attention on so-called "non-carbon forcings" —

---

[19] Editorial Board, "Palm Oil's Deceptive Lure," *New York Times* (4 May 2014), available at: http://www.nytimes.com/2014/05/05/opinion/palm-oils-deceptive-lure.html

human influences on the climate system beyond carbon dioxide. It is worth noting that in the United States, Senator James Inhofe (R-OK), a loud opponent to most action related to climate, supports action on non-carbon forcings, such as soot and other particles that come from unregulated coal plants and cook-stoves.

Similarly, there exists a wide international coalition in support of improving adaptive responses to climate — including climate change of both human and non-human origins. Food security and natural disasters are two issues where common ground can be found in a need for improved policy responses.

Carbon dioxide will remain a vexing problem because it is tied directly to the production of most of the world's energy, which in turn supports the function of the global economy. Recent experience in the United States with shale gas illustrates the virtues of innovation. Widely available, inexpensive shale gas has displaced enough coal in a remarkably short time to lead to dramatically reduced carbon dioxide emissions by the U.S.

Natural gas, a carbon-intensive fuel, is not a long-term solution to the challenge of stabilizing carbon dioxide levels in the atmosphere, but recent experience proves an essential policy point: Make clean(er) energy cheap,[20] and dirty energy will be quickly displaced. To secure cheap energy alternatives requires innovation, not just technological, but also institutional and social. Nuclear power offers the promise of large-scale carbon-free energy, but is currently expensive and controversial.

Securing innovation requires resources and a political commitment to energy as a focus of public attention,

---

[20] T. Nordhaus and M. Shellenberger, "Fast, Clean, and Cheap: Cutting Global Warming's Gordian Knot," *Harvard Law and Policy Review*, Vol. II (Jan. 2008).

much as is done in health, agriculture, and militaries. Funding could be raised through a low carbon tax—one consistent with the provisions of the iron law. A public commitment to energy innovation might be realized by recognizing the world's need for vastly more energy and the rights of billions of people to energy access commensurate with the richest around the world. An appeal to opportunity and growth will always find a stronger political constituency than demands for higher costs and limits.[21]

The need for a pragmatic way forward brings us full circle, back to where we began the exploration of disasters and climate change. If we are to respond effectively to the mounting toll from disasters, then experts will have to maintain their credibility, and policies will have to be grounded in accurate understandings.

Legitimacy and accuracy are important because the rising costs of disasters are a significant challenge for society, especially in regions that lack the political and economic resources to cope effectively with their impacts. Of course, over the long term we should all hope that a global transition to clean, cheap energy can reduce or prevent increases in the incidence and severity of extreme events due to human-caused climate change. In the short-to-medium term, though, the evidence tells us quite clearly that social change, not changes in climate, will be the dominant cause of rising disaster losses.

---

[21] For much more in this direction of thinking, I recommend G. Prins, I. Galiana, C. Green, R. Grundmann, M. Hulme, A. Korhola, F. Laird, T. Nordhaus, R. Pielke, Jr., S. Rayner, D. Sarewitz, M. Shellenberger, N. Stehr, and H. Tezuka, *The Hartwell Paper: A new direction for climate policy after the crash of 2009* (Oxford and London, UK: University of Oxford and London School of Economics, May 2010), available at: http://eprints.lse.ac.uk/27939/1/HartwellPaper_English_version.pdf

This means that, for the foreseeable future, policies that reduce disaster losses will be those that focus on increasing disaster preparedness. Linking rising disaster losses to climate change distorts the science and points us away from the policies that can be most effective in preparing for disasters. But the false link between disasters and climate change also distracts us from the many politically pragmatic and economically sensible justifications for accelerating the transition to clean, cheap energy.

In fits and starts, the real world is moving on from the climate debates of the past several decades. Those calling for action can either swim with the tide or against it — realpolitik is frustrating in that way. The legitimacy wars will of course continue, but for those interested in practical actions with consequential effects, there is a pragmatic, positive way forward.

ABOUT THE AUTHOR

Roger Pielke, Jr.
Roger Pielke, Jr. has been on the faculty of the University of Colorado since 2001 and is a Professor in the Environmental Studies Program, Director of the Center for Science and Technology Policy Research, and a Fellow of the Cooperative Institute for Research in Environmental Sciences (CIRES). His next book will be on sport in society.

ACKNOWLEDGEMENTS

This short book is the product of decades of research, collaborating, teaching and learning on this subject. There are dozens and dozens of people who I've worked with and learned from on this topic. And there are hundreds and hundreds of papers, analyses, reports and assessments which have helped to shape scientific understandings in this field. I regret that they all could not be cited here, but the reports of the IPCC and the peer-reviewed literature is only a click or two away, and I encourage anyone interested in this field to have a look.

A few people deserve special thanks for reading and commenting on earlier versions of this manuscript: John McAneney, Delphine McAneney, Dan Sarewitz, Michael Shellenberger, Ryan Crompton, Randy Dole, Marty Hoerling, Roger Pielke, Sr., and Björn Ola-Linnér. For providing data, special thanks to Ryan Crompton, Chris Landsea, Kevin Simmons, and Ryan Maue.

In addition to those mentioned above, many others have also helped me to better understand the issues associated with disasters and climate change over the years, especially including: Mickey Glantz, Laurens Bouwer, Hans von Storch, Shali Mohleji, Jessica Weinkle, Joel Gratz, Rade Musulin, Chris Landsea, Peter Höppe, Eberhard Faust, Kevin Simmons, Dan Sutter, Mary Downton, Ken Kunkel, Harold Brooks, Bill Hooke, and Stan Changnon. I've no doubt missed many others, please accept my apologies.

This book would not be possible without Dan Sarewitz, Jason Lloyd, G. Pascal Zachary, Bobbie Klein, and Ami Nacu-Schmidt, thanks. CIRES at the University of Colorado, directed by Waleed Abdalati, provides a nurturing, supportive environment for an incredibly wide range of research. I and my colleagues are fortunate to call it our institutional home.

Last to acknowledge but first in all other respects are my family—Julie, Megan, Jacob, and Calvin—who not only put up with a wonky father and husband, but are the source of endless support and inspiration.

Of course, all errors, large and small, in the text belong to me alone.

Made in the USA
Lexington, KY
07 July 2015